UNFIT FOR PURPOSE

UNFIT FOR PURPOSE

When Human Evolution Collides with the Modern World

Adam Hart

BLOOMSBURY SIGMA
LONDON · OXFORD · NEW YORK · NEW DELHI · SYDNEY

BLOOMSBURY SIGMA
Bloomsbury Publishing Plc
50 Bedford Square, London, WC1B 3DP, UK

BLOOMSBURY, BLOOMSBURY SIGMA and the Bloomsbury Sigma logo
are trademarks of Bloomsbury Publishing Plc

First published in the United Kingdom in 2020

A catalogue record for this book is available from the British Library

Library of Congress Cataloguing-in-Publication data has been applied for

ISBN: HB: 978-1-4729-7099-2; TPB: 978-1-4729-7103-6;
eBook: 978-1-4729-7101-2

2 4 6 8 10 9 7 5 3 1

Typeset by Deanta Global Publishing Services, Chennai, India
Printed and bound in Great Britain by CPI Group (UK) Ltd, Croydon CR0 4YY

Bloomsbury Sigma, Book Fifty-five

MIX
Paper from
responsible sources
FSC® C020471

To find out more about our authors and books visit www.bloomsbury.com
and sign up for our newsletters

For Donna, Emily, Sophie,
Daphne and Leo.

Contents

Introduction

Human beings are amazing. We should never forget that. Not very long ago we stared wide-eyed and barefoot at the Moon; now we can hold Moon rock in our hands. Satellites in orbit let us talk to people anywhere on the planet. We can predict the weather, explore the deepest parts of the oceans, travel at supersonic speeds, perform mind-boggling surgeries, split atoms and comprehend the very fabric of the universe itself. Day-to-day, we are no longer subservient to the vagaries of nature's rhythms to feed ourselves. No more clad in animal skins, we seldom need to venture out from our tech-heavy, luxurious homes. Slumped on our sofas, phones in hand, we can see the world, meet friends, order groceries and watch the latest movies through the greatest invention of the modern era, the internet. When we live in a world so completely of our own making, and so divorced from the natural environment, it is very easy for us to forget that beneath the blanket of modern life we are just walking, talking apes. We are amazing certainly, but we are also animals, and the twists and turns of evolution have affected us just as surely as every other living thing on Earth.

The great gulf between what we *are* as people living in the twenty-first-century world we have created and what we *were* as evolved animals in our natural environment is the central theme of this book. By examining surprisingly diverse aspects of modern life, from our diet to the rise of 'fake news', we are going to see that evolution equipped us very well for a world that, for most of us, no longer exists. Rather than helping us, our evolutionary heritage now conspires with the modern world to leave us spectacularly 'unfit for purpose'. This mismatch between the world in which we evolved and the world in which we now find ourselves has, for example, helped to create an obesity crisis (Chapter 2). It has left many of us unable to consume some important parts of the modern diet without getting ill because evolution hasn't caught up with globalisation (Chapter 3). Inside our bodies we have developed a very strained and problematic relationship with the important bacteria with which we evolved and on which our health relies (Chapter 4). Meanwhile, the stress response that evolved to save us is now killing us as we pack more and more micro-stressors into our modern lives (Chapter 5). Over the past decade we've shifted from having tens of friends to thousands and our brains simply can't handle the upgrade (Chapter 6). We evolved from a group of unusually violent mammals and we've certainly made the most of the legacy with our evolutionary tendencies to lash out creating societies riven by primitive violence (Chapter 7). Addiction, itself a cause of violence, is a result of the 'potency' of the modern world hijacking pathways in our brain that evolved to keep us alive but now push us towards a dizzying range of dangerous delights (Chapter 8). As a social animal we have evolved to function in groups, trusting in and cooperating with other people, but social evolution has also shaped us into hopelessly gullible victims of fake news and false beliefs

(Chapter 9). Most worryingly, despite a very pressing need to solve the many environmental problems we have caused, evolution has left us selfish and without any sensible notion of the future (Chapter 10). All in all, these mismatches between our evolutionary heritage and the modern environment we have created have left us in a pretty sorry state. To find out how we got into this mess, and perhaps what we can do to get ourselves out of it, we first need to get to grips with the basics of evolution.

CHAPTER ONE

A Walking Talking Ape

Evolution: a primer

We are composed of cells that stick together like building blocks to create a living, functioning organism. There are more than 200 types of cells in humans including skin cells, liver cells, nerve cells, muscle cells, intestinal lining cells, bone cells and fat cells. These cells can group together to create tissues, which are ensembles of similar cells that collectively perform some function. For example, muscle tissue can contract and, by working against our skeleton (itself made of bone tissue), allow us to move, while lung tissue enables us to exchange oxygen and carbon dioxide with the atmosphere. The grouping of different tissues together creates complex organs, like the heart, lungs, stomach and skin, that are capable of performing all the functions we need to maintain life.

The organisation of around 30 trillion cells (see Chapter 4) into a living, breathing, fighting, feeding, fleeing, breeding, singing and dancing body relies on a tightly controlled and spectacularly diverse choreography of molecular interactions within our cells. The chemical reactions that digest our food, transform molecules and allow us to perform the functions we need to stay alive are collectively termed our metabolism. These reactions must take place at the right place and time, and at rates that are neither too fast nor too slow. To achieve this organisation and control, our cells contain complex networks of protein microtubules and fatty membranes, a host of enzymes and an array of tiny structures called organelles.

DNA is at the centre of all this complexity. This famous double-helix molecule is found within the nucleus, a large organelle found in most of our cells. A DNA molecule looks like a twisted ladder with each 'rung' on the ladder being formed from a pair of so-called bases, one on each 'upright', that reach across and weakly bond with each other. Each base is effectively a letter in the genetic code that directs our cells to make proteins. There are four bases: adenine, thymine, guanine and cytosine, often known by their initials A, T, G and C. The sequence of bases along the DNA molecule is a code that is used by structures in our cells called ribosomes to make molecules called proteins. Proteins have a huge range of functions in our body. They can be structural, like the collagen in our skin or the keratin of our hair; they are the bulk of our muscles; and, as hormones and enzymes, they help to control the many chemical reactions throughout our body. Proteins are built up from long chains of small chemical building blocks called amino acids. These amino acids (there are more than 20 to choose from) join together in a very specific sequence, which determines the properties and the function of the resulting protein. This precise sequence of amino acids is dictated by the order of letters within the DNA molecule. Each amino acid in the protein is coded for by a specific sequence of three bases, called a codon. All of this complexity and organisation arose through the process of evolution.

When we think of human evolution we are usually preoccupied with the more recent stages of our development as a species: characteristics like walking on two legs (bipedalism), our large brain and the evolution of language. These features are important, indeed defining, but we must also understand that most other aspects of our biology have been subject to evolution, in some cases occurring very early in the beginning stages of life itself. The wonderfully complex processes underpinning our metabolism, our nervous system and the synthesis of the proteins that build and control our bodies all evolved long before our ability to walk upright and talk about the weather. For example, the sequence of 10 reactions that comprise a process in our cells called glycolysis occurs in the same way in us, in pandas, in yeast and in bacteria. Glycolysis is the first step in converting the sugar glucose (from our food) into adenosine triphosphate (ATP), a kind of wonder-molecule that delivers energy to cells. This series of reactions and the enzymes that control it evolved long before skeletons and muscles, eyes and ears, logic and language. Likewise, the way our muscles contract at a molecular level involves biochemical pathways and molecules that can also be found in sea anemones, starfish and earthworms. The mechanics of our skeleton, how we digest food, the ways we respond to stress and a whole host of other processes happening at the level of cells, tissues and organs are fundamentally identical to those found in other animals, and indeed in some cases plants, fungi, single-celled organisms and bacteria. That is not to say that the last couple of million years of our evolution haven't resulted in more than a few tweaks here and there, but for the most part the basics were laid down a very long time before we came onto the scene.

Biochemical pathways, enzymes, different cell types, bird bills, fish scales, hedgehog prickles and behaviour are all examples of adaptations: characteristics (or traits, to use the evolutionary term) that benefit those individuals possessing them. The development of such traits is usually moderated by specific sequences of DNA that we call genes. We have two copies of each gene, inheriting one from our mother via her egg and the other from our father through the DNA in his sperm. A sequence of DNA 'letters' making up a gene (or genes) results in some property of the individual bearing those genes. This property might result in some individuals surviving and breeding better than others who lack the gene. The fact that genes can be passed on from one generation to the next, and will be present in more offspring of successful parents, is what leads to evolution.

Evolution implies change and biological evolution is, at its most fundamental, a change in gene frequency over time. In other words, evolution occurs when genes become more or less common. Different forms of genes arise through mutations, changes in the sequences of DNA bases in the eggs or sperm of parents that are subsequently passed on to offspring. Such changes occur when mistakes happen in the process of copying DNA to make eggs and sperm, and mutations happen naturally at a low but appreciable rate. Many mutations have negative effects, but every so often a mutation may arise that provides some new way of doing things, some advantage to the offspring that inherit it. For example, perhaps a mutation slightly alters the enzyme for which it codes and the new version of this enzyme is just a little more effective in digesting food or, as we will see in Chapter 8, much more effective in breaking down alcohol.

One of the main ways in which gene frequencies can change, and therefore that evolution can occur, is natural selection. To understand how natural selection results in evolution let's imagine a gene exists for some adaptation that causes individuals that have it to do 'better' than those that do not. We can define 'better' in many ways: perhaps individuals with our imaginary gene can gain an advantage in finding food, evading a predator or securing a mate. In evolutionary biology though, these abilities are all just proxies for the only currency that really counts: the ability to leave more offspring in the next generation than others manage. The relative ability to leave offspring in the next generation is what evolutionary biologists call fitness, although being 'fitter' is more complex than simply having more offspring than your neighbour. For example, a red deer female in good condition and capable of producing plenty of high-quality milk might achieve greater fitness by having a son rather than a daughter. Her excellent condition allows her son to be well fed, and by growing large and strong he stands a good chance of becoming a dominant male controlling the reproduction of a large harem of females. By having a son, the mother gains more grand-offspring and accrues greater fitness than if she had a daughter. Likewise, a nesting bird that lays a large number of eggs may end up with far more offspring leaving the nest than her neighbour with a more cautious egg-laying strategy. However, the fact that the ambitious mother has been run ragged feeding all of her chicks means that she may not have enough reserves to live through a cold winter to breed again next spring. Her death reduces her future fitness to zero. What is more, her many offspring may also not be in good enough condition to survive the winter. By balancing current reproductive output with the chance to breed again next year, and by having a smaller number of offspring that can each receive higher quality care, the mother with fewer offspring this year might actually achieve greater fitness over her lifetime.

Let's assume that, all other things being equal, a gene leads to higher fitness for those that are lucky enough to have a copy of it. These individuals will have more offspring as a consequence of having that gene. Because these offspring have a good chance of inheriting the gene from their successful parents the gene will likely increase in frequency in the next generation. If the environment in which these individuals live, which includes physical factors such as temperature as well as biological interactions with predators, prey and parasites, doesn't change, then it is quite likely that this new generation will also benefit from the gene they inherited. These offspring will go on to have their own offspring that will also have a chance of inheriting further copies of the gene. The inescapable, beautiful, logical result of these bouts of reproduction and natural selection across generations is that genes that confer a benefit to the individuals bearing them tend to increase in frequency. If on the other hand the selection environment changes, then genes (and for many traits we are usually talking about multiple genes) that code for particular adaptations may no longer be as beneficial to their bearers. In these circumstances we will see a reduction in the frequency of less beneficial genes, and a consequent reduction in the number of individuals with the adaptation that they code for. In other words, through natural selection, gene frequencies change and evolution occurs. This process happens whether you are a human drinking a martini on a luxury yacht or a bacterium digesting dog mess.

The evolution of modern humans

Natural selection can lead to the evolution of adaptations by selecting those individuals with genes coding for traits that increase survival and reproduction, as we have already seen. Evolution can also lead to the formation of different species through processes collectively known as speciation.

Only rarely will all the individuals that make up a species be able to interbreed freely with each other. An example would be a species found only on a small island. Most species, including humans, exist in different populations of individuals that tend to breed among themselves. The birds in your garden might be the same species as the birds in the garden of someone 1,000 miles away but in practice these two groups of birds will probably not breed with each other, even though, in principle, they could. Sometimes migration of individuals to and from different populations allows for interbreeding and gene flow between populations, but overall most individuals are likely to breed with members of their species that live close by.

Different environmental pressures acting on populations living in different locations, or being active at a different time of the day or year, can cause selection for different adaptations in those populations. Perhaps in one valley where a species of mouse is found there is a different predator eating mice than is found in the rest of the mouse species' range. The presence of this unusual predator has selected for more nocturnal behaviour than is usually found in the mice because mice that prefer to come out to forage when it is dark survive far better than those favouring daylight. The valley population was already perhaps somewhat geographically isolated from others of the same species. Now, the population becomes further isolated biologically because even nocturnal individuals that disperse outside the valley are unlikely to be able to meet and mate with the non-nocturnal individuals present in the rest of the species' range. Over generations, further genetic differences can accrue as the nocturnal population evolves under the influence of the selection pressures of its new nocturnal niche. We might see the evolution of better low-light vision for example, or slightly different dentition to favour food more readily found at night. These genetic differences further increase the relative isolation of this population with respect to the rest of the species. Eventually the valley population becomes sufficiently distinct physically, ecologically and behaviourally for us to identify it as a new species.

To understand how our species, *Homo sapiens*, is affected by the evolution of adaptations unique to our species and by those adaptations that are common to a much wider group of organisms requires us to examine our evolutionary history. The problem is that our evolutionary history extends all the way back to the very start of life. Clearly, we are going to need to draw the line somewhere. To keep things manageable and relevant, I'm going to constrain this history to the emergence of what paleoanthropologists call anatomically modern humans. When necessary, for example when discussing the evolution of the mechanisms that lead to stress (Chapter 5), violence (Chapter 7) or addiction (Chapter 8), it might be necessary to delve a little deeper.

To define a species, and to classify it clearly and unambiguously as different from any other species, we need to define the characteristics that are shared by members of that species but not present in any other group, at least not in the same complete combination. You might think that defining a 'human' is neither especially difficult nor important. We never find ourselves in the position of wondering whether any given collection of flesh and bones in front of us is a human being or not and we don't need to tally up anatomical features to be sure. In terms of an effect on our everyday lives, having a firmly nailed-down definition of what makes us part of the species *Homo sapiens* is indeed pretty inconsequential. However, by understanding the very specific features that make us 'human' we are much better armed to consider the role that those evolved features have in our everyday lives.

One problem with defining any species is that individuals vary. Humans are no different and any definition of our own species must account for this variation. Some of the variation in humans arises more or less solely from genetic differences between populations. Such differences might be highly apparent: the differences in skin colour between a white European

and an African-American for example, or between the
stocky stature and physique of an Inuit indigenous to
Nunavut in northern Canada and the slimmer form of
the Sān peoples of Namibia. You don't need to travel to
see other clear genetic differences between humans.
The most profound variation is between males and
females and these differences are also genetic, caused by
the presence of a Y chromosome (male) or an X
chromosome (female) in the sperm that fertilised the
egg from which we developed. Other genetic differences
might be less apparent and related to disease susceptibility
or metabolism. A good example of a metabolic difference
between populations that is genetic is the way our
bodies deal with lactose in milk and gluten in wheat
(Chapter 3).

While some variation is certainly genetic, other
sources of variation are solely environmental. A good
example of a highly variable characteristic that is determined purely by our environment is the primary way
that we communicate. People growing up in different
places and cultures speak different languages. While
our linguistic ability is genetic and related to physical
structures like the larynx (the 'voice box') and tongue,
and to specific regions of our brain, the precise form our
language takes is related to what language we hear as we
develop from birth.

Yet more variation can arise from environment and
genes interacting. Adult height, for example, has a genetic
component, but without an adequate diet during
childhood an individual bearing genes promoting tallness
is unlikely to achieve their full potential height. The
balance of genetics and environment, of 'nature' and

'nurture', is of great interest both scientifically and philosophically. It is also a topic of great relevance to this book, because evolution is fundamentally a genetic process. If we are to discuss the role that evolution has had in shaping us, and the mismatch between our evolved selves and the modern world, we are inevitably making assumptions that whatever features we are discussing have a genetic basis. As we will see throughout this book, despite being extremely seductive, such assumptions are not always easy to substantiate. So 'nature' is going to be critically important to our discussions, but so too is 'nurture', since it is the interaction between our present-day environment and our genes that leads to many of the mismatches that we see.

Currently, there is only one living species in our genus (*Homo*), but in the geologically very recent past there were other *Homo* species wandering around that are now extinct, most notably *Homo neanderthalensis*. We know from genetic studies of modern humans and of well-preserved material from recent fossil records that Neanderthals and early humans interbred. We have also found other fossil material that indicates there were more *Homo* species in the mix, including *Homo naledi* in South Africa and *Homo floresiensis*, the famous 'hobbit' found in Flores, Indonesia. These other hominins, as members of our genus are known, were probably also interbreeding with each other in the run-up to the emergence of anatomically modern *Homo sapiens*. The piecing together of the emergence of anatomically modern humans and the reconstruction of our recent evolutionary past is a difficult, controversial and often fast-moving field; the discovery of a bone fragment here or a tooth there usually results in a big media splash and yet another rethink of sequences and dates. In short, it's a bit of a mess and it's likely to remain so for some time. Despite this complexity and confusion over some of the details of our recent evolution, we can reasonably safely say that archaic *Homo sapiens* were first striding around the Earth about 300,000 years ago and that fully anatomically modern humans arose around 160,000 years ago. The balance of evidence currently points to an African origin with a subsequent spread of our species into Europe, Asia and beyond. At the same time that our earliest modern human forebears were emerging in Africa, other *Homo* species were evident in more far-flung locations. Footprints of *Homo heidelbergensis* have been found in southern Italy for example, and Neanderthals were widespread across Europe and southern Asia as long ago as 400,000 years.

What makes us different

The anatomically modern humans that emerged around 160,000 years ago have numerous shared characteristics that differ from other hominins. First and foremost, the skull is a very distinctive shape. Unlike all other hominins (and indeed primates generally) we have a very weakly developed supraorbital arch, or brow ridge. A brow ridge is, however, present in remains of earlier *Homo sapiens* (those dating from between 160,000 and 300,000 years ago) and can also be seen in some humans alive today. Aboriginal peoples of Australia, for example, often exhibit a brow ridge, but it is very different from the ridge found in Neanderthals and other archaic members of the genus

Homo. Where present, modern brow ridges are not generally complete across the eyes, and it is usually only the central section that is visible. The function of the brow ridge is to bolster the skull, providing structural reinforcement against the powerful forces generated by working the bottom jaw when chewing. It was a feature that was lost late in human evolution and its loss is related to another prominent feature of our skulls, a steep forehead.

Our almost vertical foreheads are in stark contrast to the sloping foreheads of other hominins. A surprisingly complex part of our anatomy, the forehead has a series of three muscles that allow us to display a range of expressions including quizzical and surprised. That the forehead is a useful billboard for our emotions, at least for those shunning the Botox needle, is a great example of evolution co-opting something that arose for an entirely different purpose, in this case related to housing our large and complex brain. To accommodate such an organ required a different type of space from the one that the sloping forehead of other hominins provided. Before we start getting too carried away about our big brains, though, it is worth noting that Neanderthals actually had a slightly larger brain. The general consensus from studies of their skulls suggests that Neanderthals were able to accommodate such a large brain despite possessing a sloping forehead by having skulls that extended lengthways as opposed to upwards, creating the more spherical skull shape that we have. It is this 'rugby ball' versus 'football' design that reveals the key advance in the modern human brain. Our brain has enlarged frontal lobes and it is this part of the brain that is heavily involved in all the cognitive processes that we tend to associate with modern humans: decision-making, planning, creativity, social behaviour and abstract thought. A big frontal lobe extends the brain forwards and upwards and to accommodate that you need a steep vertical forehead. That big frontal lobe was a driving force in our success and the sometimes subtle effects that it has on our fit to the modern world will be seen throughout this book.

Exemplified by the loss of the brow ridge, the overall structure of our skull is more delicate and fragile than the thicker and more robust skulls of other hominins. We have more delicate lower jaws with smaller teeth, especially our canines and incisors. We also have a pronounced chin that combined with our relatively short jaws and steep foreheads makes our proportionally smaller faces almost vertical. This gracility extends to the rest of our skeleton, with relatively longer limbs and thinner bones than any other hominin yet discovered. The extra length of our arms relative to our bodies does not come from equal (or isometric) growth of the limb bones but via disproportionate (or allometric) growth of those long bones nearest the hands (ulna and radius) and the feet (tibia and fibula). Overall, our build differs from the robust build of other hominins and it is thought that these more slender proportions are an adaptation to life in warm tropical regions. As a body becomes less stocky and more slender it reduces its surface area relative to its volume and loses more heat, a useful adaptation for an active animal in very hot regions.

Interestingly, stockier proportions are apparent in populations of humans who evolved to live in the cold northern Polar Regions today where, conversely to living in the tropics, heat retention is a good thing. This is sometimes said to be an example of Bergmann's rule, one formulation of which states that individuals within a species (especially warm-blooded mammals and birds) tend to be larger at higher latitudes.* For humans though, it is not body size but body shape that is the key player here and Allen's rule, that body shape tends to be more rounded and compact in colder climates, is a better fit.[†] That we can understand human evolution and variation using the same rules that we apply to other species further reinforces the point that we are just animals, subject to the same pressures of selection and processes of evolution as all other organisms.

An obvious feature of hominins compared to most animals is bipedalism, the ability to walk upright on two legs as the preferred gait. Bipedalism is not unique to hominins, as even the most casual observation of

---

* Bergmann's rule is named after German biologist Carl Bergmann (1814–1865). Bergmann noticed that within groups of related species there was a tendency for those species in colder regions to be larger. This is now often applied to patterns found within species, where individuals tend to be larger and more robust in colder climates. Like many rules in biology it doesn't always hold true, but nonetheless it does describe the patterns found in a diverse range of mammals and birds.

† Named after the American zoologist Joel Asaph Allen (1838–1921), this rule states that animals adapted for colder climates tend to have shorter limbs and appendages (such as ears) than animals adapted for warmer regions. It is better supported for individuals within a species, including us, than for patterns between species where other factors (including Bergmann's rule) tend to become more important.

birds reveals, but it is unusual in mammals. Kangaroos and wallabies, for example, are bipedal and so too are the springhares, a curious group of African rodents superficially resembling rabbits. A number of mammals, including bears and many primates, can walk on two legs for a short time and some tree-dwelling primates, notably the gibbons, exclusively use a bipedal gait when walking on the ground. Outside of the mammals and birds, basilisk lizards can run on their back legs at such a rate that they can achieve the feat of walking on water (a feat that earns them the name Jesus Christ lizards), but this can only be maintained for short periods. The elegant, permanent bipedalism of humans is only possible because of the configuration of our backs, pelvis and supporting ligaments, tendons and muscles. To walk upright in any sort of convincing fashion it is necessary to take long strides without falling over. This is far from easy and the problem of bipedalism still vexes many of those working in the robotics industry although (pun intended) great steps forward have been taken and we can now see the development of some very effective bipedal robots. Key to solving the problem of bipedalism is finding a way to balance the upper body over the legs in a manner that allows the legs to move and develop a stride pattern without toppling. We achieve this mainly through a curved spine and changes to our pelvis that together make our skeletons very distinctive.

So far all the features we have explored that define humans have been skeletal. Clearly, our skeletons and related features are of great importance in defining our physical abilities and limitations, but modern humans cannot be defined by bones alone. Complex tools, the use of fire, the development of art, music, language and abstract thought are all human cultural features that have resulted from our large, complex and powerful brain. Given that we are often interested in comparing modern humans with other hominins, a focus on skeletons, hard parts that fossilise and persist until their discovery, is unsurprising. Increasingly though we are taking more interest in whatever products of hominin culture we are able to discern from the sparse archaeological and fossil record. The findings are often surprising. Burial rituals equating to what we would surely regard as a funeral involving fire and the seemingly symbolic use of animal horns can be inferred from Neanderthal sites, for example.[2] The discovery of *Homo naledi*, a hominin with an intriguing combination of modern and primitive features (notably a rather small brain), in a cave in South Africa in 2013 revealed perhaps the most surprising insight into possible non-human hominin culture. The remains were pushed far into a chamber accessed through a series of climbs, crawls and squeezes; the evidence points to the body having been interred there deliberately. Such behaviour is important because it implies the capacity for abstract thought and symbolism, traits usually ascribed only to *Homo sapiens*.

The fact is we are still learning what truly makes us human and we are still unravelling the complexities of our evolutionary past. This book is about mismatches between that evolutionary past and the environment we have created, and exploring these mismatches is going to require some serious intellectual juggling. The story of why we are unfit for purpose in the modern world is a story of genetics, natural selection, evolution, biogeography, genome analysis, biochemistry, sexual selection, archaeology, psychology, sociology, politics and much more besides. So, pull up a chair, grab some calorie-dense snacks and let's start with a very 'big' mismatch; why the modern world conspires with our evolutionary heritage to make us fat.

CHAPTER TWO
Chewing the Fat

Evolution is bloody marvellous. It is, quite simply, the means by which we understand much of biology. The staggeringly talented British immunologist Peter Medawar, described by the evolutionary biologist and essayist Stephen Jay Gould as 'The cleverest man I have ever known', said that 'For a biologist, the alternative to thinking in evolutionary terms is not to think at all.' Gould himself was no stranger to lauding evolution, stating that it is 'one of the half-dozen shattering ideas that science has developed to overturn past hopes and assumptions, and to enlighten our current thoughts'. So while others may have put it more eloquently than me, I think that, when faced with the mind-bending diversity of adaptation and life on Earth, 'bloody marvellous' does just as good a job.

For all its greatness, there is a problem with looking at life through the lens of evolution. We can sense this peril lurking in one of the most famous quotes of all about evolution, the title of an influential essay by Ukrainian-American geneticist and evolutionary biologist Theodosius Dobzhansky: 'Nothing in Biology Makes Sense Except in the Light of Evolution'. It is such a wonderful title, bordering on an arrogance virtually impossible to ignore. But if you think about it, taken at face value it simply isn't true.

The problem with the brilliantly sweeping statement of Dobzhansky is that not *everything* we see in biology, not every feature of animals, plants, fungi, single-celled protists or bacteria, is necessarily an evolved adaptation. In other words, not every feature has come about because there is a genetic component underlying that feature that confers an advantage to the bearer of those genes that translates into increased fitness and resultant increase in gene frequency. The problem is that the lens of evolution is so powerful and its explanatory power so great that viewing the world through it can lead to a tendency to see the hand of natural selection and evolution in absolutely everything. Adaptationism is the scientific position that ascribes an evolved adaptive origin for many of the features that we can document in organisms and it is a wonderful approach. But adaptationism can be taken too far. Take a simple physical feature of humans, our belly button. This universal structure of all mammals nurtured in their mother's womb via an umbilical cord is wonderfully variant in form, and unparalleled in its ability to take a piercing and go septic, but it is not an evolved characteristic. There has been no selection for the belly button. Belly buttons are not a genetic characteristic that have conferred such an advantage to those that have them that the genes coding for them have swept through the population to fixation (when all individuals have the genes) in just a few generations. Belly buttons are merely the attachment point of the umbilical cord connecting you to your mother's placenta. They are basically a scar. Thanks Theodosius, but belly buttons make total sense without the light of evolution. That said, the umbilical cord that produces the belly button is an evolved adaptation, so it can all get a little complex to disentangle.

'Armchair adaptationism' is sitting at a distance and concocting evolutionary Just So stories, named for the Rudyard Kipling short stories like 'How the Leopard Got His Spots'. It is always a temptation to interpret every feature that we can see in organisms as an evolved characteristic, and that is never more of a problem than when we turn the lens of evolution on ourselves. Applying the beautifully simple hammer of selection logic is just so tempting and satisfying that we can end up seeing nails everywhere. So, why this cautionary preamble? Well, in this chapter, I'm going to explore the complex, emotive and biologically difficult topic of obesity. As we'll see, the route to understanding why we are fat is littered with tempting nails dying to be hammered down with evolutionary arguments. We will need to remember that just because an evolutionary Just So story is seductive and 'obvious', it doesn't mean it's true.

Seductively thrifty genes

I'll explore this statement more scientifically later but for now let's just stick to a tabloid level of analysis and accept that humans are 'getting fat'. The well-repeated evolutionary argument as to why we are so fat these days is that we are famine-adapted creatures that find ourselves living in times of feast. The story goes like this. Prior to the invention and spread of agriculture about 12,000 years ago in the so-called Neolithic Revolution (see Chapter 3), we were operating on a largely subsistence basis. This period, the Palaeolithic (the Stone Age), saw our ancestors and other hominins with whom we shared common ancestry living as hunter-gatherers. We'll get into more details of their diet and environment shortly, but as a first approximation we can assume that these early humans would have been relatively non-selective omnivores. Plant products, including high-protein and calorie-dense nuts, vitamin- and carbohydrate-rich berries and starchy root tubers would have been particularly prized and, judging from archaeological evidence, meat also featured on our early menu. Meat would likely have been acquired opportunistically by grabbing whatever didn't move fast enough. Small mammals, birds, eggs, insects and shellfish would have been consumed with gusto when the opportunity presented itself and the flesh of larger animals could have been scavenged from carcasses. We also developed increasingly technological hunting methods using tools, communication and social organisation to bring down large prey. In times of plenty, perhaps when a host of berries were in season, frogs were croaking in the lakes and game was plentiful, it is easy to imagine our ancestors full-bellied and satisfied. However, the hunting would not always have been good. Animals can be remarkably hard to find and kill (that's evolution for you), while plant bonanzas are highly seasonal. Full bellies might be easy to imagine but hard times and hunger would never have been far away.

We are capable of ingesting far more calories in the food we eat than we need to survive. In times of plenty we could easily have had an excess of calories available to us. To get through hard times of famine it makes sense to be able to store the surplus energy available during times of feast. This is the precise reason why honey bees make honey; it is a food store laid down when nectar is highly available in the summer and can be converted to honey to last through the winter when no flowers are in bloom. While honey bees have evolved an external food store, we have evolved an internal food store, or fat as it is known. Our history of feast and famine, so the story goes, has led to *Homo sapiens* being highly adept at laying down fat stores in times of abundance to survive the leaner times ahead. In the modern world, and most especially in More Economically Developed Countries (MEDC), we no longer go through periods of feast and famine but instead are surrounded near-constantly with calorie-dense food. The upshot is that we live in times of absolute feast but are physiologically ever-prepared for a future famine that will almost certainly never come, but just in case it does, our bodies lay down fat and so we balloon. This evolutionary argument has become known as the thrifty gene hypothesis (TGH).

Since it was proposed back in the 1960s, the thrifty gene hypothesis has gained considerable ground in the public sphere and it has become the go-to evolutionary explanation in the media for why obesity is a crisis in the modern world. Indeed, such is the power of this hypothesis that obesity has become something of a figurehead for the notion that those of us in MEDCs have created a world for which we are not adapted. It also suggests a very simple remedy: eat a diet more like our ancestors and you shouldn't get fat.

As simple as it sounds, the evolutionary cause of, and evolutionarily informed solution for, the obesity problem subdivides into a number of different and sometimes complex steps that need to be supported with evidence if we are to understand and solve the obesity problem with evolutionary thinking. We need to identify what fat is, what obesity actually means, how obesity is related to other conditions and whether this is really a feature of the modern world or whether obesity was around long before we started 'going large' on our burger and fries. Then, we need to dig down (in some cases literally) to examine what our ancestors ate and what their environment was like compared to ours, especially with respect to famine (a central feature of the TGH). If we are to blame an evolutionary mismatch for current trends in obesity, then we are going to have to explore the selection pressures our ancestors faced and compare them with those that are exerted by the environment in which we live today. Once that ground is laid, we need to bring out the big guns; we need to look for the genetic evidence for a 'thrifty' gene or genes because if evolution is to blame then it must have left its signature in our genome. Finally, the evolutionary-inspired claims that backdating our diets to those of our paleolithic forebears will solve the obesity problems need to hold up to scrutiny. Paleo diets and caveman nutritional plans need to be more than just marketing gimmicks. It's a house of cards because if the evidence for any one of those steps is lacking, it starts to cast considerable doubt on the whole enterprise. So, let's start with the simplest question: what is fat?

What is fat?

In Chapter 1 I touched on the fact that our bodies are composed of cells that can be organised collectively to form tissues that perform a particular function. One such cell-collective is adipose tissue and it consists of cells called adipocytes, also known as lipocytes (fats being biochemically part of the family of substances called lipids). More plainly, we can call these cells 'fat cells'. Fat cells are wonderful at storing fat and they collect together to form the adipose tissue you can feel when you grab a chunk of soft yielding flesh.

Adipose tissue actually comes in two types, brown and white. Brown adipose tissue (BAT) is found in almost all mammals and in humans it tends to accumulate around the kidneys and neck, between the shoulder blades and along the spinal cord. BAT is involved in producing heat by the process known as non-shivering thermogenesis. Shivering produces heat (thermogenesis) as a by-product of the involuntary contraction of our muscles. Non-shivering thermogenesis allows the chemical energy in fat to be converted to heat without the need for movement via a clever metabolic pathway present in BAT cells. BAT is medically intriguing, being associated with, among other things, bone density, bone health and increased longevity. As a consequence of all these functions BAT is metabolically active with a good blood supply, and the fat cells themselves have small, accessible droplets of fat stored within them. They also have plentiful mitochondria, the 'powerhouses' of cells that are required to fuel BAT cells in their complex (and still not fully understood) role.

As interesting as BAT is, it is white adipose tissue (WAT) that makes us 'fat'. BAT is brown because of the plentiful mitochondria within the cells and rich blood capillary networks supplying them, whereas WAT is white because it doesn't need to do anything especially interesting beyond storing energy. BAT cells have small droplets of fat stored within them but the fat cells of WAT have a lovely big space, or vacuole, right in the middle. This vacuole can fill up with energy-rich fat, and when full the vacuole pushes all the other cell contents (like the nucleus) tight up against the cell membrane.

WAT is present all over our bodies. Subcutaneous WAT fat forms just under the skin, whereas visceral WAT fat accumulates deep in the abdomen and surrounds our organs. Subcutaneous fat leads to beer bellies, love handles, amply cushioned buttocks, large thighs, a jowly face and flabby arms. It tends to be the fat we can see and the fat we worry about aesthetically, but in terms of our health the evidence suggests that subcutaneous fat is not causing the main problems. The really big negative health implications of fat storage come from the visceral fat we cannot see piling up around our internal organs. A high amount of visceral fat is a major player in the development of type 2 diabetes, which is one of the biggest medical problems associated with obesity. Type 2 diabetes is an increasingly common condition that causes the level of glucose (a type of sugar) in the blood to elevate because we become resistant to the effects of insulin, the hormone that regulates our blood sugar levels. Symptoms include excessive thirst, frequent urination, increased hunger, tiredness and, a symptom that isn't talked about so much, an itching around the penis or vagina. These are relatively minor short-term symptoms compared to the very serious long-term implications of type 2 diabetes, which include heart disease, strokes and impaired blood flow that can lead to blindness, kidney failure and, in extreme cases, foot, hand and limb amputations. Other solid links, both direct and indirect, have been established between obesity and heart disease, strokes, certain types of cancer and psychological problems including depression and low self-esteem. To this we can add the pain caused from excessive joint wear, especially of the knee, and sleep disorders including sleep apnea (an interruption to normal breathing while asleep) and excessive snoring. As if these weren't enough to persuade you that obesity is a bad thing, obese people are 25 times more likely to report problems with their sex lives, whether it be lower libido, erectile dysfunction or issues linked to body image and performance anxieties. It's undeniable: being obese is bad.

Obesity and diabetes

As well as being one of the major health problems associated with obesity, type 2 diabetes plays an interesting part in our consideration of the importance of past evolution on our present state. It was in attempting to account for the rise of type 2 diabetes that the thrifty gene hypothesis was first postulated in 1962 by geneticist James Neel. Neel was an influential character in the development of an evolutionary understanding of human genetics and disease, although his work was not without controversy. In 2000, author Patrick Tierney in his book *Darkness in El Dorado* laid out a series of very serious accusations primarily against Neel and anthropologist Napoleon Chagnon, although other researchers also came under fire. The most serious of the accusations made by Tierney was that Neel had initiated a measles epidemic in 1968 among the Yanomami of the Amazon. The charge was that Neel had deliberately used a hazardous version of the measles vaccine in order to test theories on human evolution, the existence of 'leadership genes' and the evolution of infectious diseases in humans. According to Tierney, Neel had used the Edmonston B type of measles vaccine even though a safer and cheaper version was available. The motivation, so the claim went, was that the Edmonston B vaccine produced antibodies that would allow Neel to compare European and Yanomami immune responses. Tierney asserted that the use of this vaccine to fulfil a scientific purpose rather than using a safer alternative led to a measles epidemic. The *Darkness in El Dorado* incident as it related to Neel certainly illustrates how emotive human evolutionary biology can be, but in fact the measles epidemic claim, and indeed a great many other claims in the book, were later categorically refuted.[1]

What can't be refuted quite so easily is the very clear link between obesity and type 2 diabetes. Paradoxically, given that both are disadvantageous, obesity and diabetes are both highly heritable. Heritability suggests a genetic basis and a genetic basis suggests evolution arising from selection, but how can selection operate to promote the increase in frequency of disadvantageous genes? It was in grappling with this conundrum that Neel hit upon the notion that insulin resistance (a precursor of type 2 diabetes) could promote the efficient storage of fat that would allow individuals to survive periods of famine. Thus, a disadvantage (a tendency to promote diabetes) is associated with an advantage during human history when food was scarce and the pros outweighed the cons. It was quickly realised that the link with diabetes was not actually required for selection of thrifty genes as long as they promoted survival against the current environmental background. Fat deposition was a good thing for survival against a background of regular famines and so was born the evolutionary explanation for the heritability (and so the genetic basis) of getting fat.

Getting to grips with BMI

Feeling like you're carrying too much timber, being able to grab a little more than you used to around the belly and having the feeling that your clothes are shrinking are all good everyday signs that you are likely to be putting on weight. Medically, obesity is defined a little more precisely. The globally accepted measure for whether you are obese or not is the Body Mass Index or BMI. Adopted by medical professionals and organisations around the world including the World Health Organisation (WHO), BMI is calculated by dividing weight in kilograms by height in metres multiplied by itself, or squared. So, I weigh 83kg (183lb) and I am 1.88m (6ft 2in) tall, which means my BMI is 83 divided by (1.88 x 1.88) = 23.5, give or take. Under current definitions, a BMI between 20 and 25 is 'normal' (phew), between 25 and 30 is 'overweight' and above 30 is 'obese'. BMI then increases in jumps of 5, with every jump representing a different and progressively more severe level of obesity, passing through severely obese (35) to morbidly obese (45) and ending (currently) on hyper obese (60+).

BMI has a mathematical logic that is hard to ignore. Taller people are generally heavier and by dividing weight by height you can allow for that and compare more meaningfully between different people. In short, having a higher BMI than someone else means that you are carrying more weight for your height than they are. There are two main issues with BMI, the first of which the more mathematically astute among you may have noticed. When you scale up a three-dimensional object (like a human body) along one 'axis' or dimension (like height) you also have to scale up in the other two dimensions (width and depth). Imagine a cube with sides of 1cm (roughly ½in). If you double its height, you must also double its width and length to keep it as a cube. The volume (which equates to its weight) of the original cube is 1 x 1 x 1 = 1cm³, but the volume of the doubled cube is now 2 x 2 x 2 = 8cm³. A more logical formulation of BMI, therefore, would seem to be weight divided by height cubed rather than squared. This was pointed out by Professor Nick Trefethen in a beautifully concise letter to *The Economist* in 2013:

> SIR – The body-mass index that you (and the National Health Service) count on to assess obesity is a bizarre measure. We live in a three-dimensional world, yet the BMI is defined as weight divided by height squared. It was invented in the 1840s [by a Belgian statistician and sociologist called Adolphe Quetelet], before calculators, when a formula had to be very simple to be usable. As a consequence of this ill-founded definition, millions of short people think they are thinner than they are, and millions of tall people think they are fatter.[2]

Trefethen explores this issue further and, acknowledging that people don't scale in quite such a perfect way as cubes do (when we grow taller we don't grow wider and deeper in the exact same proportions), suggests going midway, raising our height to the power of 2.5 rather than 2 (squared) or 3 (cubed). The wonderfully named Corpulence Index (CI) does actually use the cube of height but it never really gained any traction and, for now at least, BMI is here to stay.

Another issue with BMI, and one that many people seem to cling to without always taking much account of the evidence in front of them, is that more muscular people tend to be heavier. The argument goes that muscle is denser than fat and so if you are carrying a lot of muscle mass then you may well be heavier for your height than you 'should' be according to BMI calculations. Consequently, you could have a BMI that suggests you are overweight when in fact it is your well-developed musculature that tips you over the edge. If you're thinking of using this argument then please do take some time to look in your trophy cabinet of Olympic medals for rowing and weightlifting, or at your bodybuilding schedule; if either of them are full then maybe you have a point. For most people though, the 'it's not fat, it's muscle' argument is not likely to be a winner. BMI also does not tell us anything about the relative proportion of subcutaneous or visceral fat, so it is possible to be carrying potentially dangerous levels of visceral fat but not fall into the 25+ overweight category. Likewise of course, it is possible that your BMI could be 25+ but that your fat is all carried around the buttocks and thighs with fewer health implications.

I guess what I am saying is that BMI may have a few issues when it comes to specific individuals, but arguing over the fine details doesn't detract from its usefulness when we apply it at a population level. BMI is an effective and accepted measure that allows us to compare between different populations and between and among populations over time, and when we do that very clear patterns emerge. We are fat and getting fatter.

Global obesity patterns

Worldwide obesity has nearly tripled since 1975, and in 2016 39 per cent of adults (aged 18 or over) were

overweight (BMI 25–30) and 13 per cent were obese
(BMI 30+). What that means is that about one in seven
adults on planet Earth right now is obese and two out of
every five adults are medically overweight. These numbers
are only going up, at least for now. Largely related to this
rise in obesity, we have also seen a global increase in
diabetes (from 108 million in 1980 to 422 million in
2014).[3] But if we are to blame the rise in obesity and
the subsequent decline in our health on a mismatch
between our evolutionary past and the developed world
in which many of us now live, we might expect there
to be a relationship between obesity and some measure
of 'development'. To work that out requires more
exploration of the data.

Examining the global distribution of those with a
BMI of 30 ('obese') or more is an interesting first step in
understanding the rise of obesity, not least because it
yields some rather surprising findings. For example, the
top eight obese nations are all Pacific Island nations,
headed by Nauru with a scale-busting 61 per cent of
people being obese. The first non-Pacific Island nation is
Kuwait at number 9 (38 per cent obese) and then comes
the USA, flying the flag for fatness at number 10 with 36
per cent of its population being obese. Arab nations
feature highly in the back half of the top 20, representing
seven of the next eight nations, from Saudi Arabia to
UAE (the eighth nation being Turkey at number 15). In
the bottom nine of the table, with obesity levels less than
5 per cent, there are poorer nations such as Ethiopia,
Eritrea and Nepal but also Japan, a wealthy and highly
developed country. Simply 'eyeballing' the numbers
isn't enough to determine whether being part of the

developed world leads to an evolutionary mismatch and therefore obesity. To do that, we need to take a more considered approach.

It is difficult to find a single measure of nation 'development' that everyone feels happy with and so rather like BMI we have to accept that there is a degree of give and take in our deliberations. The United Nations Human Development Index (HDI) for example is a statistical measure that takes into account factors like life expectancy and education, whereas Gross National Income (GNI) or Gross Domestic Product (GDP) are blunter economic measures of productivity. A more nuanced but less snappily named economic measure, 'GDP (at purchasing power parity) per capita', accounts for the cost of living and inflation rates of different countries and provides an arguably better measure of 'lifestyle' and development. A number of analyses have used a variety of different measures and in fact found much the same thing. Despite 'noise' in the data, with some spectacularly obese poorer nations (see below) and some very trim richer nations (e.g. Japan), there are relationships between most measures of economic development and obesity.[4] We should be wary of the old adage that correlation does not imply causation, and we should also be very mindful that global obesity cannot be simply waved away with a trite explanation that as we get rich we get access to food aplenty and get fat. But equally clearly, something is going on.

The Pacific Obesity Cluster
The standout anomaly when considering the relationship between obesity and development is the prevalence of

extreme obesity in the relatively poor Pacific Island nations, occupying the top eight positions of the global obesity league table. Surely this 'Pacific Obesity Cluster' presents a problem for the overall idea that development relates to obesity despite the overall trend? As we'll see, in fact the clustering of these nations supports the underlying hypothesis and, rather neatly, provides us with an excellent example of the power of recent human evolution to drive mismatches with the modern environment we have created.

We are using development as a proxy measure for access to calorie-dense foods and arguably also a more sedentary lifestyle, which is another driver of obesity. The Pacific Island nations actually support the idea that access to calorie-dense food is the primary driver of obesity, but the major difference between those nations and the wealthier nations that follow them in the league table is that the supply of calorie-dense food is driven not by wealth but by relative poverty. In Tonga (number 5) for example, the diet changed in the mid-twentieth century from fish, root vegetables and coconuts to imported offcuts of meat. Fatty 'turkey tails' from the US and 'mutton flaps' from New Zealand were cheap and rapidly became popular. Mutton flaps are the low-quality end of sheep ribs and are a staggering 40 per cent fat. Such calorie-laden delicacies are frequently the only meat available and many lay the blame for present-day obesity in these nations firmly at their door. There are likely other cultural factors at play in the South Pacific, such as the historical association of large size with wealth, beauty and power, but the main factor causing present-day obesity in these nations is ready access to surplus calories. Genetic research though reveals that there are relatively recent evolutionary factors at play in the South Pacific that have led to greatly elevated tendencies among these islanders to lay down fat stores. When coupled with a recent environmental change resulting in ready access to calorie-dense foods like mutton flaps, this genetic background conspires to produce the most obese nations in the world.[5,6]

The story of Samoa

The selection pressure that caused the evolution of 'super-obesity' goes a long way to support the thrifty gene hypothesis, at least in the South Pacific. To find out more we must visit the sixth most obese nation on Earth, the beautiful Pacific Island nation of Samoa.

Samoa, like the other Pacific Island nations that lead the global fat league, was settled by people that most likely travelled from Taiwan through the Philippines and eastern Indonesia. This consensus 'out-of-Taiwan' model of expansion and settlement of the Pacific is supported by evidence from linguistics (the study of language similarities and differences), genetics and archaeology, and the expansion likely happened around 3,000 years ago. The settlement of Samoa and other island nations would clearly have required considerable ingenuity in boat building, including the development of double-hulled voyaging canoes, and skill in open ocean navigation. It would also have required physical resilience, toughness and the ability to cope with serious food shortages. Boats have limited capacity to carry stores, fresh food would not last long and voyages were long in terms of both distance and time. There would be a clear advantage to those individuals that were able to lay down fat stores more effectively when food was available, perhaps before the voyage or during stop-offs along the way, and to make good use of those fat stores during long periods of on-board famine. Indeed, the long voyages necessary to settle the islands of the South Pacific would seem to provide an ideal selection regime for genetically 'thrifty' individuals. But with the caution I advised earlier about the perils of armchair adaptation and evolutionary seduction, we need to examine the genetic evidence to see if our Just So story holds up.

One method of working out whether there is a genetic link to a particular condition, trait or characteristic is to perform a Genome-Wide Association Study, or GWAS. In principle, GWAS are straightforward. You identify a group of people that have the trait you are interested in, in this case people who are obese, and a group of people who don't. You then genotype all the individuals in both groups and compare the resulting long list of genetic 'letters' (the sequence of DNA bases we met in Chapter 1), looking for consistent differences between the groups. Genetic characteristics shared by those in the group with the condition and lacking in the group without the condition are the smoking gun required to ascribe a genetic association to the condition. Of course, in practice GWAS can be very difficult. To be robust, studies require a great deal of genetic know-how and some heavy-duty statistical analysis using tools developed in the incredibly important field of bioinformatics. You also need to control for factors like sex and age in your study population as well as the underlying geographical and ethnic background of participants. These factors can all produce associations that are caused not by a genetic association with the trait of interest but by genetic associations between the participants in each group. Nonetheless, these issues can be resolved and GWAS have been applied medically with great success since the start of this century. Developing through a series of landmark studies of myocardial infarctions in 2002 and age-related macular degeneration (a cause of vision loss in older people) in 2005, the most notable early success came with a 2007 study that revealed, through the study of 17,000 participants, a genetic underpinning for seven diseases including type 1 and 2 diabetes, coronary heart disease and rheumatoid arthritis.

Even allowing for a more generous BMI cut-off of 26 to allow for differences in skeletal build and muscle mass between Europeans and Samoans, 80 per cent of men and 91 per cent of women in Samoa were overweight in 2010. As I discussed in Chapter 1, there could be a solely environmental reason for this (mutton flaps streaming in because of a relatively poorer economy), a solely genetic explanation (actually impossible because you cannot get fat without food, which is always a product of your environment), or an interaction between environment and genes. In this case, the interaction would be between mutton flaps and some putative 'thrifty genes'. In fact, even before the onset of the modern-day obesity epidemic, the prevalence of obesity in Samoans was greater than in other populations. This strongly suggests that modern-day environmental changes in diet may not be the only factor in the rise of obesity and lends some support to there being an inherent, genetic component to obesity in these islands. Further indication that there might be a genetic influence on BMI in Samoa, and strong evidence at that, is the fact that there is an estimated heritability (the component of a trait determined by genes and capable of being passed on to children) for BMI of 41 per cent. Something is clearly going on in the population of Samoa and a GWAS is the best way to find out exactly what.

The year 2016 saw the publication of a GWAS focused on high BMIs in Samoa. Studying more than 3,000 Samoans, the research team were able to identify a specific variant of a gene called CREBRF that was very strongly associated with BMI. CREBRF is a gene identified widely across vertebrates and expressed (in other words, read and used by cells to make a protein) in many different tissues, including adipose tissue. Studies suggest CREBRF has a fundamental role to play in the function of our cells; these studies have also implicated the proteins coded for by CREBRF and related genes in the metabolism of fat cells. Overall, the background evidence suggests that this gene could well have a role to play in fat storage and obesity. Those Samoans with a higher BMI were more likely to have the variant of the CREBRF gene, which was also associated with body fat percentage, abdominal circumference and hip circum-ference. CREBRF seems like a good candidate for a thrifty gene.

Nearly one in two Samoans carry at least one copy of the variant CREBRF gene associated with high BMI in Samoa. In contrast, the variant is unobserved or extremely rare in other populations (one in 18,300 Europeans and one in 2,100 East Asians). That the Samoa variant is rare in other populations is not perhaps surprising. Modelling using a computer tool called Polyphen-2 predicts the impact of genetic mutations on the structure and function of proteins, and this approach suggests that changes to the specific part of the gene impacted in the Samoa study are highly likely to be damaging and are therefore unlikely to spread. To unravel this conundrum, the researchers studied the effects of the genetic variant using what is known as a 'cell model'. Cell models use established 'lines' of specific cells that can be grown in the lab and used to investigate how cells function. The cell model the researchers used was 3T3-L1, a cell line of adipose cells established in 1962 from the embryo tissue of a Swiss albino mouse and widely used to study the cellular basis of fat metabolism. Using this cell model they found that the variant gene decreased energy use and increased fat storage in these cells, which is surely the very definition of thrifty from a fat-cell perspective. Taken together with the history of the Samoan people and the population genetics in Samoa and beyond, the study presents a very convincing evidence-base for the CREBRF variant found in Samoa being a thrifty gene.

In an interesting twist to the original Neel formulation of the thrifty gene hypothesis, as well as promoting fat storage the Samoan variant also seems to protect against type 2 diabetes. Carrying the variant of the gene which causes a 1.3-fold increase in the risk of obesity actually reduces the risk of type 2 diabetes 1.6-fold, which is certainly a counter-intuitive finding given the well-established link between obesity and type 2 diabetes. That the gene provides protection against one of the key diseases associated with the condition it promotes is likely linked to the effect that it has on metabolism at a cellular level and could certainly explain why the gene is still around and at such a high frequency in the Samoan population.

Further evidence for South Pacific genetic thriftiness has come from additional studies of obesity in people from this region. A study of Māori and Pacific people living in New Zealand replicated the findings of the Samoa study, revealing the same genetic variant having the same association with higher BMI and reduced risk of type 2 diabetes.[7] What is more, the magnitude of the risk associated with the variant was also similar. So, related people with a similar recent evolutionary history share a rare genetic variant that promotes fat storage. Couple that with a shared heritage of long sea voyages and onboard famines that likely killed off the thin and it feels like the TGH is well and truly over the line. But things are not all they seem.

Problems with the thrifty gene hypothesis

The Samoan study does provide excellent support for the thrifty gene hypothesis, but when it came along in 2016 the TGH was losing ground despite being widely parroted in the media and supporting a number of dietary fads. The problem for the TGH was not in establishing the truth behind the current environment because we quite clearly have better access to calorie-dense food now than we did in the past. I can nip to one of the many supermarkets within a five-minute walk of where I work, buy a kilogram of sugar and wash it down with a wine glass full of molten lard for less than £1. Though not the nicest meal (and eating that much sugar is certainly not recommended), I could, in no time at all and with very little financial outlay, consume around 6,200 calories, or enough to keep me going for a good two and a half days. The problems for the TGH actually came in establishing the feast–and–famine selection environment in human history that would be required to cause the evolution of thrifty genes, and then finding definitive evidence of those genes in our genome. These are not insignificant problems by any means.

A particularly major problem with the TGH is that if there have been a great many famines over the course of human history, and a greater risk of death to leaner individuals lacking the fat stores to make it through, then the force of selection on genes promoting fat storage would have been massive. Massive selection pressures promoting certain genes should push those genes to fixation: the point at which everyone (or virtually everyone) carries the beneficial gene. Such evolutionary change can happen surprisingly rapidly, even with quite modest selection pressures. John Speakman, a major critic of the TGH, has done the mathematics and the resulting numbers are not good news for the hypothesis. The calculation goes like this. First, Speakman assumes that a mutation in some gene linked to fat storage arises and that this mutated version of the gene promotes obesity in some way. Carrying this gene gives an individual only a 0.5 per cent selective advantage over the rest of the population, in other words they are just 0.5 per cent more likely to survive a famine than someone carrying the normal version of the gene (the 'wild type' as geneticists would call it). Speakman then proposes a schedule of famine events, such that there is one famine every 150 years. With these conditions in place, none of which seem especially challenging or controversial, a mutation that was present in a single individual would spread to fixation and be found in the entire population after 6,000 selection events (famines) or, in this case, 900,000 years. Granted that extends beyond the period of anatomically modern humans, but it is well within the period we might consider to be our recent evolutionary history. We should, says Speakman, all be obese against such a selection background, but we are not. The variation in obesity and tendency to deposit fat among humans works against the TGH, because a beneficial thrifty gene should have spread to fixation.

Another issue with the TGH arises from the pattern of mortality that we typically find in famines; in other words, who tends to die when famine strikes. The TGH relies on body composition being the prime factor driving mortality, with lean individuals more likely to die than those that are fatter. Providing that the fatter individuals are fatter as a consequence of some genetic mechanism promoting body fat, then a thrifty gene can spread. The problem is that overall there seems to be very little evidence supporting the idea that fatter people are more like to survive famines than thinner people. As Speakman cautions though, the lack of evidence for an effect does not necessarily mean that the effect is lacking and we have no studies comparing BMI and the likelihood of death in a population before and after famine.

In fact the primary factor affecting mortality in famines is not body composition but age. Famines tend to kill the young (<10) and the old (rather ungenerously defined as 40+). Selective death affecting the old will have little or no effect on evolution. Deaths that occur after individuals have produced offspring are invisible to natural selection, unless perhaps we are dealing with a social species where older non-breeding individuals have some genetically underpinned and significant part to play in the survival of closely related kin. An example would be grandmothers providing care for grandchildren and promoting their survival and subsequent ability to have offspring (see Chapter 3). At the other end of the age-scale, the deaths of children are also unlikely to play much of a part in the evolution of thrifty genes. Since childhood obesity is a relatively very recent phenomenon, Speakman's argument is that famine-related mortality cannot have been biased towards leaner children in the past. In essence, all children were lean during early human famines and were equally likely to live or die regardless of whether they carried a thrifty gene. Perhaps most damning of all, where data do exist on the overall mortality imposed by famines, the numbers are not especially high. A study of 190 identified famine events occurring in the UK across 2,000 years indicated that famines occurred every 10 years. However, examination of the likely mortality effect of those famines showed that only one famine every 100 years produced a significant increase in mortality above the baseline mortality for the previous 10 years. The overall conclusion is that famines are not especially common and when they do occur they do not result in particularly high increases in mortality. In summary, famines, the primary selection factor in driving the evolution of putative thrifty genes, do not happen very often, do not have a large effect when they do and the individuals killed by famine in a population are not those that are consistent with the hypothesis.

Speakman explores a number of different selection environments and pressures and through his calculations concludes that 'If the thrifty gene idea is correct, we should all have inherited advantageous mutations in thrifty genes, and if these mutations cause obesity, as the hypothesis suggests, we should all be obese.' Of course, you cannot get fat on genes alone; it is still essential to have an environment of calorific plenty in order for that genetic background to be expressed. It could be argued that the fact we are not all obese is not necessarily clinching evidence against the TGH, because we can, at least notionally, control what we eat even if we are surrounded by cakes, crisps, cookies and chocolate. Furthermore, if we look at the UK then we are certainly in a period when the majority of people are overweight, with 62 per cent of adults having a BMI greater than 25 in 2014. It is predicted that as many as one-third of adults in the UK could be obese by 2020 and that proportion has already been comfortably breached in the USA. Nonetheless, Speakman makes a strong case by stating that if selection for thrifty genes were a long-standing feature of our evolution then thrifty genes should be widespread and, crucially, that fatter people should fare better in times of famine.[8]

It is nearly always difficult to disentangle the different effects of genes and environment on a particular charac-teristic, and this is also the case if we are looking for thrifty genes that were important in our early evolution. Global obesity statistics combine a great number of intermeshed environmental effects, including recent history, culture, economic development and food availability, that will interact with genes that might be common to all humans, or might be, as we have seen in the South Pacific nations, more or less contained within specific populations. If thrifty genes were essential in our early survival then we might be able to see their effect in populations of modern humans that live in environments and have lifestyles more in keeping with our evolutionary past, or at the very least have lifestyles that are not 'modern', characterised by readily available high-sugar, fatty, processed foods. In fact, the effect of a thrifty gene should be easy to spot. The crucial advantage of having such a gene is that between periods of famine, individuals carrying the gene can get fat, building up stores that allow them to survive the next food scarcity. So, the logic goes, we should see people getting fat between famines. The reality is that we don't.

The year 1816 became known as 'the year without a summer'. In April 1815 Mount Tambora, a volcano in Indonesia, erupted or, to put it more accurately according to contemporary reports, exploded. It was the largest observed eruption in recorded history and the sound of the mountain tearing itself asunder was heard a staggering 2,600km (over 1,600 miles) away. Ash filled the skies and resulted in a global climatic phenomenon known as a volcanic winter. Particles of ash and clouds of sulphur dioxide gas spewed into the atmosphere. Reacting with water in the atmosphere, the sulphur dioxide formed sulphate aerosols that together with the ash particles obscured the sun, reduced the heat reaching the surface of the Earth and caused temperatures to decrease across the world by 0.4–0.7°C. The drop, although a good deal less than a degree, was sufficient to provoke severe effects on the climate in many places.

In North America, for example, the atmospheric pollutants from the eruption caused a persistent fog to form, and the lowered temperatures resulted in unseasonal frosts in May and June of 1816. Further frosts were reported in July and August. The effects on agricultural production that year were devastating and the consequent crop failures have been called 'the last great subsistence crisis in the western world', promoting near-famine conditions.[9]

If we treat 1816 as a famine event then, if the argument for thrifty genes holds up, we would expect the population to lay down fat from that point on in thrifty preparation. However, historical data from the US reveal that in the late 1890s, a full 80 years later, the level of obesity in the USA was still only 3 per cent,[10] well below the level we see now and certainly not consistent with thrifty genes squirrelling away fat as an adaptive response to famine. It seems much more consistent with a small number of people having access to plentiful food while the majority of people ate a simpler, more calorie-controlled diet without the fatty processed food and abundant sugar that we find in more recent times.

As well as looking back through time, if we want to remove the influence of the modern world on BMI we can look to contemporary populations that are relatively unaffected by the modern world. Hunter-gatherer societies are now rather unusual and tend to be found in the less accessible and more remote parts of the world such as the San people of the Kalahari, the Hadza of Tanzania and the Jarawas of the Andaman Islands in the Indian Ocean. Collecting food predominantly by foraging, such people are effectively living a similar lifestyle to our pre-agricultural ancestors 12,000 years or more ago. It is important to say here that these people are not Stone Age humans. They are modern humans with their own recent history of evolution. My point here is merely that their lifestyle is not what we would call modern. As well as hunter-gatherers, there are many more communities around the world that practise subsistence agriculture, relying purely on what they themselves produce. Hunter-gatherers and more particularly people relying on subsistence agriculture might expect to exhibit strong evidence for thrifty genes, since food shortage and interruptions in availability cannot be smoothed out by reliance on wider society or alternative means; if you don't find it or grow it, you don't eat it. However, studies on such communities during non-famine conditions (when thrifty genes should be promoting fat deposition) consistently find BMI from 17.5 to 21, which is at the very lean end of normal.

The thrifty late hypothesis

That existing hunter-gatherers are lean, well fed and do not seem to show any sign of thrifty genes, at least from the studies that have been carried out, is interesting because it points us towards an elaboration of the TGH that could help us to get around some of the issues that Speakman raises. Around 12,000 years ago the development of agriculture saw a major shift in our feeding behaviour. No longer were we solely tied to the vagaries of the environment and natural food availability; but equally, the development of more centralised, larger agricultural societies made us potentially more prone to famines. The development of agriculture saw more mouths to feed and with all our eggs in fewer baskets, both literally and metaphorically, the impact of droughts, crop failures and animal diseases might be felt ever more keenly by a population that was now too large and lacking in knowledge to revert suddenly to a foraging existence. To account for this seismic shift in the lifestyle of most modern humans, the 'thrifty late hypothesis' suggests that intense pressure for thrifty, fat-storing genes would only have occurred in the last 12,000 years or so. 'Thrifty late' explains why hunter-gatherer populations do not become fat between famine events, because they have not been under the same agriculture-induced selection pressures as other populations. By having a shorter period of more intense selection, the thrifty late hypothesis also seems to go a long way in explaining why we see such a variance in fat deposition between people in response to modern diets (some people gain weight more readily than others) and why there are still a large number of unfixed variants of genes related to obesity. Another supporting element was inserted into the adaptive thrifty gene hypothesis by British nutritional scientist Andrew Prentice in 2008. Prentice agrees that famines are likely only to have been a significant factor in relatively recent, agricultural human history but argues that it is reduced fertility rather than increased mortality that is their primary effect.[11]

Obesity: are there benefits?

If the TGH, or amendments like 'thrifty late', are correct then we really should be able to detect evidence that at least some of the genes associated with obesity are providing an adaptive advantage. After all, it is the counter-intuitive fact that obesity is clearly bad, and yet equally clearly obesity has a genetic component, that is really at the heart of the TGH. We can argue about selection pressures, the magnitude of famines and the impact of agriculture all we like, but the TGH really does demand that there are some genes that cause obesity that also benefit those who carry them. We can investigate that by looking for signals of selection in the human genome at those places (termed 'genetic loci') that are associated with obesity.

The 1,000 Genomes Project is the ideal tool for probing the genome (our DNA) to find out more about its selection history. The project was launched in 2008, and by 2012 1,092 human genomes had been sequenced (their long sequences of DNA bases determined) and made freely available for scientists to study. Using these genome sequences, Guanlin Wang worked with Speakman to look at the 115 genetic loci that were known in 2016 to be associated with obesity. By examining the genome data in detail, it was possible for them to determine that only nine of these loci showed evidence of the positive selection that would be a signature of the thrifty hypothesis. What is more, of those nine loci, five (so more than half) actually involved positive selection for leanness rather than obesity.[12]

So, by the time the Samoan study came along, the TGH was not faring well; as a human-wide global explanation for the recent rise in obesity it arguably still isn't. The prevalence of obesity in South Pacific nations and the identification of a thrifty gene associated with obesity and protective against diabetes in those populations is excellent evidence that the TGH has some merit, but looking across the wider human population the evidence is, like our hunter–gatherer ancestors, rather lean. We cannot, it seems, firmly pin the obesity epidemic on a mismatch between our thrifty past and our calorie-rich present. A much simpler explanation is that like other mammals we have the ability to convert excess fat into adipose tissues, and these days it is simply far too easy for many of us to eat far too much.

If not thrifty then drifty?

Despite some clear issues with the thrifty gene hypothesis, we are still left with an intriguing genetic association with obesity and a relatively high heritability of obesity, both of which suggest that something interesting is going on. The prevalent counter-hypothesis has become known as the 'drifty gene hypothesis' (DGH). Like the TGH, DGH is an evolutionary explanation, but unlike the TGH it is not an *adaptive* hypothesis. To understand how the DGH might explain obesity, we first need to understand a process called genetic drift.

Evolution, remember, is the change in gene frequency over time and in general we are concerned with the evolution of adaptive traits through natural selection. We may also be interested in other adaptive selection mechanisms, including sexual selection (the peacock's tail, for example) or kin selection (honey bee workers committing suicide for the good of the colony). However, selection for and against favourable and unfavourable traits is not the only way that genes change in frequency over time.

Consider a bag full of ten marshmallows, five of them white and five of them pink. Now imagine that these marshmallows can reproduce, their offspring are exactly the same colour as them and there is no real difference between the two colours in terms of the likelihood of surviving and having offspring. To start filling a second bag with our 'next generation' we simply pick out, without looking, a marshmallow. If it is pink then we put a new pink marshmallow into the 'next generation' bag and we put the 'parent' marshmallow back in the original bag. Then we pick again, without looking, and put our second 'next generation' marshmallow into the second bag, returning the parent to the first bag as before. This carries on until we have ten marshmallows in each bag.

Assuming we are unbiased in our selection, it feels like we should end up with five white and five pink marshmallows in the next generation because we are picking from an evenly proportioned population. In reality though, we would only get five white and five pink in our next generation an average of one in four times. By sheer chance we could actually end up with a new population that was entirely white (one out of every 1,024 times we did it) or entirely pink (the same chance as entirely white). Odds of one in just over 1,000 indicate events that could happen but are pretty unlikely, so let's say we get some intermediate value, like seven white and three pink (120 chances out of 1,024). White would have increased in frequency (evolution) but without any adaptive advantage. The next time we sample (to create a third generation) we could, again just by chance, end up with an even more biased population. Indeed, at some point the proportion of white marshmallows might have changed towards the point where it has gone to fixation and pink no longer exists. This non-adaptive change in gene frequency occurring because of the random sampling of individuals is known as genetic drift.

The possible role of genetic drift in obesity was presented by John Speakman at the 2007 Obesity Society meeting in New Orleans as part of the presidential debate, with Andrew Prentice. Prentice, you'll recall, is the nutritionist we met earlier, who proposed that reduced fertility through famine was more important than increased mortality through famine in the evolution of thrifty genes. Speakman's argument, which he has developed considerably since that first outing, combines models of how we control our weight and fat storage with an ecological perspective of human evolution. Central to the argument is the threat of predation.

There are a number of theoretical approaches, or models, that have been proposed to understand how our bodies control 'fatness'. The three basic models are the set point, the settling point and the dual intervention point, and there is plenty of debate and experimental work developing and investigating each of them.

1. The set point model proposes that there is some, you guessed it, 'set point' of fatness and that the body is trying to regulate fat stores up or down to reach this ideal. It works well as a model to explain weight gain and weight loss in rodents, and some evidence supports it in humans, but the obesity epidemic suggests that perhaps a set point model isn't the full picture. If we have a set point for fatness then why are so many of us so far away from it, and why is the set point

seemingly so variable across even reasonably closely related populations?

2. The settling point model proposes that the set point can move depending on nutritional conditions and the environment. So, in an environment with plenty of food and relatively little energy expenditure through physical exercise our body weight 'settles' to a new higher point. The settling point model seems more consistent, initially, with what we see in human populations, but observations of people on diets and taking part in controlled feeding experiments are actually more consistent with the set point model. With aspects of both models supported or refuted by studies on mice and humans, neither seems a satisfactory explanation for fatness in humans.

3. The dual intervention point model tries to account for the seemingly conflicting observations that there is evidence supporting a set point but that environmental factors also seem to be able to alter body fatness. The dual intervention model proposes that as well as an upper limit controlling our maximum weight, there is also a lower limit preventing us from becoming dangerously thin. The lower limit, or intervention point, is the weight at which our body intervenes to prevent starvation and activates fat storage. In a sense, this lower limit is a kind of 'thrifty' limit, a saving regime imposed when times are tough and the belt has, literally, been tightened too far. The upper limit is far more interesting. The upper limit is proposed to be linked to our ability to avoid predators.

The logic behind the predation-imposed upper limit, or intervention point, is simple. If you are fat you are less able to escape the claws and teeth of animals pursuing you for food. It is frankly very hard to argue against the logic here. Given just how bad being eaten is in terms of survival, future reproduction and lifetime fitness, it certainly makes sense that avoiding it by remaining a lean, mean, running-away machine could have been a vital component of evolutionary history. I have a reasonable amount of personal experience in examining the insides of various African antelopes and a very striking thing about them is the almost complete absence of fat. These prey animals are the very definition of lean and while that does put them at risk during periods of food scarcity (such as droughts), that is a long-term, 'cross that bridge when they get to it' concern (see Chapter 10 for more on this). In the short term, that leanness translates into a remarkable ability to run away that is pretty useful when you live next to lions and leopards.

Speakman uses similar evolutionary thinking combined with ideas of genetic drift to explain why the upper intervention point has 'drifted' upwards over time in modern humans. Around 2–4 million years ago our hominin ancestors, little more than small primates living in open savannah habitats, were fair game for a suite of predators no longer around today. Those early hominins were the antelopes of their day, prey animals that hid, cowered and ran away. Sabre-toothed cats belonging to the wonderfully named genus *Dinofelis* ('terrible cat') and the fearsome *Megantereon cultridens* roamed these plains, and evidence from gnawing marks on bones suggests early hominins were most certainly on their menu. Against such an environmental background of teeth and claws there would be a very strong selection pressure on early hominins to remain lean and to have a tightly constrained upper intervention point.[13] Being fat would in all likelihood mean being dead. But then, along came *Homo erectus*...

Larger in body size, *Homo erectus* made use of fire and tools and was a more sophisticated hominin in every way. We also know that *H. erectus* formed larger social groups than previous species. It is thought by some that this social behaviour, which would have involved evolutionary changes to the brain, may well have been an adaptation against predation, although sociality could also increase hunting success and allow for bigger prey to be taken. A bit later, around 1–2 million years ago, we know from the fossil record that a number of large-bodied carnivores went extinct in east Africa. Suddenly, predation wasn't such a big concern for members of the genus *Homo*. Suddenly, you didn't need to be quite so thin, quite so lean and quite so ready to run. Suddenly, if you had a tendency to put on a bit of timber, it might not matter and you might make it to an age when you could have some little *Homo*-lings, who might inherit your tendencies towards a more ample waistline.

With a relaxation of selection pressure on that upper intervention point, mutation and genetic drift could have become very important factors in the evolution of obesity. Adding to their contribution is the fact that early populations were small. Small populations tend to exacerbate the sampling effects we saw earlier with marshmallows, which can be ironed out to some extent when you have more individuals to sample. Overall,

Speakman contends that, unconstrained by predation, the upper intervention point has drifted around and has resulted in the large diversity of upper intervention points (i.e. the large diversity in obesity) that we find in modern-day humans.

Neither the thrifty gene nor the drifty gene hypotheses have yet gained universal approval and a number of scientists continue to develop different arguments and counter-arguments, amendments and counter-amendments, tweaks and refinements to both hypotheses.[11,14] Human populations are highly variable in all kinds of ways and the discovery of thrifty genes in some populations but not others might simply be a reflection of that diversity. It may also be that drift has been more important in some populations than in others. Overall though, the simple idea that we are fat because we are storing up for a forthcoming famine, as seductive as it undoubtedly is, doesn't currently stand up to full scrutiny across all populations. Even if the drifty gene hypothesis is correct, at least for some populations, then we still get to blame evolution for getting fat, albeit in a far less satisfying way. Rather than relatively recent ancestors being metabolically good boy scouts, we have far more distant ancestors drifting towards obesity because they didn't need to run as much. Ancestral laziness, not thriftiness, may be the reason our clothes are shrinking.

An evolution-inspired solution?
If we stand back from the fine print and accept that evolution has had a role in determining how we store fat (and that is really beyond dispute), then the big question

is, can we counter the current global obesity crisis simply by adopting the diet of our ancestors? The Palaeolithic diet, also known as the paleo diet, the Stone Age diet or the caveman diet, is the idea that shunning the foods of the modern world and eating only those foods that were available to our ancestors will align our diet with our evolved bodies and prevent us from getting fat.[15] Whether this is a good plan or not rests on several things: knowing what our 'caveperson' diet actually was; whether adopting that diet will help us to lose weight, the goal of most people choosing a diet plan; whether the diet is healthy; and whether the diet is sustainable in the long-term so that people can stick to it and keep the weight off.

Determining what our ancestors ate requires us to examine a wealth of different types of archaeological evidence. We know from bones showing tool marks and scorching from cooking for example that our ancestors ate meat. Hunting of many species is also depicted in cave art. We also know that they enjoyed seafood when they could, with great piles of shells stacked up on beaches where our ancestors feasted. Likewise, fruits, berries and nuts were eaten, eggs were consumed where they could be collected and even honey would have appeared in the diet.[16] Given their availability and the fact that they are eaten widely across the world today, it also seems likely that insects would have made at least the odd starter here and there.

We can delve deeper into ancient nutrition by analysing teeth and bones using a technique called stable isotope analysis.[17] This technique relies on the fact that nitrogen and carbon both exist in different forms, called isotopes. The nucleus of an atom of carbon always has six positively charged protons and this 'atomic number' (6) is what defines the atom as carbon. Atomic nuclei also have neutrons, sub-atomic particles that carry no charge and have the same mass as a proton. The number of neutrons in the nucleus of carbon can vary and this variation results in heavier and lighter forms, or isotopes, of carbon. The same applies to other elements like nitrogen, which has an atomic number of 7 (seven protons) but can have six or more neutrons, with the most common number being seven.

Carbon actually has 15 isotopes but only two of them, carbon 12 (six protons and six neutrons) and carbon 13 (six protons and seven neutrons) are stable for any meaningful length of time. Carbon 14, used in radio-carbon dating, decays slowly into nitrogen with half of it decaying in 5,700 years (its half-life). It is this decay that allows us to use it as a kind of atomic clock to date organic material. Nitrogen has 16 isotopes but again only two are stable, nitrogen 14 and the less common nitrogen 15. The ratio of the different stable isotopes of carbon and nitrogen present in a sample of bone produce an isotopic signature that enable us to make inferences about the diet that helped to form that bone. Different plants, for example, produce different ratios of carbon isotopes, with diets rich in subtropical grasses producing a ratio more skewed to carbon 13 than diets richer in products derived from trees and temperate-zone plants. Nitrogen isotope ratios can reveal the extent of meat-eating as well as indicating whether a diet had a large component of seafood.

Archaeological and isotopic evidence allow us to piece together a reasonable picture of the sort of foods that were consumed in different places and at different times. Notable components of a modern diet missing from the Palaeolithic diet include grains in any great number and dairy foods, both of which are products of the later development of agriculture (see Chapter 3). Of course, the paleo diet also lacked the processed foods so familiar to us today, so no cornflakes, bread, sausage rolls, ready meals, cakes, biscuits, sweets or fizzy drinks.

To go from the list of food types and some general overview of proportions of different broad types of food (plants versus meat, seafood versus 'land-food') that we can derive from archaeological evidence to an actual day-to-day schedule that we can adopt in the modern world is something of a leap. Cave paintings might tell us that our ancestors ate meat and collected honey, but they are pretty light on the sort of detail we need for a diet plan. To help us make that leap we can look to existing hunter-gatherer societies that are largely isolated from the influence of a modern diet for more clues. This approach, combining archaeological evidence with contemporary observations, underpins arguably the most influential book in the area of caveperson nutrition, *The Paleo Diet*, published in 2002 and written by nutritionist and exercise physiologist Loren Cordain.

Cordain didn't come up with the idea of the paleo diet, but he was perhaps the most important force in propelling the concept into the mainstream. In fact, the nutritional principles underlying the book mostly derived from a paper published in 1985 by Stanley Boyd Eaton and Melvin Konner that laid out the foundations for the paleo diet. Their paper, entitled 'Paleolithic Nutrition', appeared in the prestigious medical journal

New England Journal of Medicine and in it Eaton and Konner explore archaeological dietary evidence, lay out their findings of a review of contemporary hunter-gatherer nutrition and identify the mismatch between the diet of the past and modern human nutrition with its links to obesity. Their final line sets the stage for what would become a very profitable dietary trend. As they put it, 'The diet of our remote ancestors may be a reference standard for modern human nutrition and a model for defense against certain "diseases of affluence".'[18]

Eaton and Konner developed this idea further, publishing the book *The Paleolithic Prescription* in 1988 with Marjorie Shostak, an American anthropologist who conducted fieldwork with the !Kung San bush people, a group of hunter-gatherers living in the Kalahari desert. This book greatly promoted the evolutionary 'discordance hypothesis' (as they called the mismatch between our evolved heritage and the modern world), and brought the concept of evolutionary mismatch into the mainstream. It was against this background that Cordain's book was published and the paleo diet has been firmly in the public eye ever since. Indeed, a quick search on Amazon reveals more than 10,000 books on the topic and a Google search yields more than 94 million hits. Despite competition from the carnivore diet, the keto diet, the Dubrow diet, the noom diet and even the biblically inspired shepherd diet, the paleo diet is still very much alive and thriving.

There is no official paleo diet. It is more of a dietary philosophy, with different authors proposing different variations in how to interpret the basic tenet and providing different daily suggestions and recipes. Whether this is to provide some clear water in a very crowded marketplace is not for me to say, but taken overall, some clear paleo-diet trends emerge. Paleo-diet advocates reject dairy, grains, sugar, legumes, processed oils and salt, since such foods would not have been available in Palaeolithic times, while embracing fruit, vegetables, nuts, meat and fish. Cordain recommended that 55 per cent of daily calories come from an equal mix of lean meat and fish and the remainder from an equal mix of fruit, vegetables, nuts and seeds. Other authors tweak these figures but most remain in that ball park. Basically, it seems to come down to looking at the food in front of you and asking, if it was 20,000 years ago and you were standing naked on the savannah with a spear, or on the shoreline with a pointed stick, would this food have been possible? A pepperoni pizza washed down with a glass of milk scores a big no, a piece of meat with some vegetables scores a yes. Of course, in answering this question we do need to overlook the fact that most of the vegetables we identify today have evolved from anthropogenic artificial selection in recent times and bear little resemblance to 'paleo veg'.

The really big question though is not whether your fish and chips counts as 'paleo' (as long as you don't fry it in oil or coat your fish in flour) but whether the diet really works. In 2017 Jonathan Obert and colleagues published a review of the scientific and medical literature associated with four weight strategies, one of which was the paleo diet. If you are looking to lose weight rapidly then their results are encouraging. At least nine trials have shown short-term benefits of the diet that include weight loss and reduction in waist circumference. This is positive news for paleo advocates, but Obert et al. are critical, stating that trials were short in duration and 'underpowered', which is essentially saying that there were too few people being tested.

A 'gold standard' procedure for investigating the effects of a particular intervention or treatment is to use a randomised control trial (RCT). RCTs allocate subjects randomly to either a treatment group or a control group, with comparisons between the outcomes of the two groups providing evidence on the efficacy or otherwise of the treatment or intervention. An RCT of the paleo diet was undertaken in Sweden, the results of which were published in 2014. Seventy post-menopausal women with BMIs in excess of 27 were allocated to either a paleo diet (35 women, average BMI 32.7) or to a Nordic Nutrition Recommendations (NNR) diet (35 women, average BMI 32.6). The paleo diet was such that 30 per cent of calories consumed came from protein, 40 per cent from fat and 30 per cent from carbohydrates, and it was based on the usual suspects: lean meat, fish, eggs, vegetables, fruits, berries and nuts. Additionally, subjects were allowed avocado (a fat source), and rapeseed and olive oil to be used in food preparation and dressing. Neither of those oils, by the way, pass the 'standing naked on the savannah' test. As would be expected, dairy products, cereals, added salt, refined fats and sugar were all excluded. The NNR diet aimed for a daily calorie breakdown of 15 per cent from protein, 25–30 per cent from fat and 55–60 per cent from carbohydrates, with an emphasis on low-fat dairy products and high-fibre fruit and vegetables. Overall, both groups had a calorie intake that averaged 2,000kcal a day on the diet compared to around 2,300kcal before, so by no means a crash diet regime. In fact, the trial lasted for a full 24 months and inevitably involved some pretty big lifestyle shifts, especially for those on the paleo diet. To support them, and to ensure that they stuck to the diets and provided useful data, the subjects were given ample back-up including recipes, cooking classes and group meetings.[19] It is always worth remembering that these sorts of trials rarely take place in a clinical setting and so a degree of trust is required; if a paleo dieter sneaked a quick burger and milkshake, the researchers would never know.

After six months, both groups had lost body weight, and BMI and waist circumference were significantly lower than when they started. The paleo dieters lost an average of 6.5kg (14.3lb) of body fat in six months, which is more than double the 2.6kg (5.7lb) lost by the NNR dieters. That is an impressive result: a stone of weight in old money. Further losses occurred over the next six months but interestingly both groups of dieters hit a plateau in terms of body weight and the other measures after twelve months. That six-month lead for the paleo dieters didn't last and by 24 months both groups were level pegging.

Despite losing its early advantage, the paleo diet clearly 'works' in terms of losing weight, and works well. As well as overall weight loss, rapid early losses are beneficial because when people can see and measure the difference it tends to keep them motivated. Lest we get too starry-eyed and atavistic here though, we should consider the fact that in the long-term (two years) the paleo diet is no better for weight loss than a 'normal' calorie-restricted diet. So, in terms of losing weight on a long-term plan, the components of the diet (twenty-first century versus Stone Age) are less important than the simple measure of energy surplus. What the study confirms is the boring and predictable fact that if the 'energy in' is less than the 'energy expended' you will lose weight. The conclusion is that, from a weight-loss perspective, the paleo diet works simply because by following any of the many variants of it you are restricting your calorie intake. There is nothing 'magical' about the components of the diet, and if you were to eat double paleo-portions every day you would gain weight just as surely as if you were 'going large' down at the burger joint.

Worryingly for those thinking of going Stone Age, a number of potentially serious health implications of the diet have been identified, at least in some studies. Participants in a study of the effectiveness of the paleo diet in Australia reported significantly more episodes of diarrhoea on the diet than those following a conventional calorie-restricted regime, and 69 per cent of participants reported an increase in their grocery bills.[20] Interestingly, and certainly not to be discounted, 43 per cent of participants reported a belief that the diet was not healthy (perhaps greatly swayed by their regular and doubtlessly unpleasant toilet visits). One health risk that is commonly associated with the paleo diet is reduced bone density leading to osteoporosis, a condition especially associated with the elderly where weakened bones commonly lead to fractures. Dairy is a major source of calcium in our diet and by excluding dairy the paleo diet tends to be low in this vital bone-forming mineral (more on this in Chapter 3). Long-term studies of the effect on bone density of following the paleo diet are lacking, but paleo-diet advocates are quick to point out that calcium is present in foods other than dairy and that magnesium and vitamin D are also required for bone formation. All this is true, but when sources of magnesium are flax seeds and artichoke hearts, and calcium source examples include bone broth, oysters and figs, I can't help but question the day-to-day sustainability of such a diet for most people.

Mismatches and our evolutionary heritage

The mismatch between our twenty-first-century diet and the simpler diets of our ancestors is the best known of the evolutionary mismatches being blamed for our current woes, but as we've seen, the famine-adaptation thrifty gene hypothesis is not widely supported by the evidence, at least across the human population overall. The drifty gene hypothesis is gaining ground and is an evolutionary explanation, but it is a more complex and less satisfying formulation. Drifty genes is a non-adaptive hypothesis that takes us far deeper into our ancestry and involves more wide-ranging ecological factors than we might have expected. Whichever hypothesis wins out, and in reality we might be looking at a complex mix of both and others we have yet to devise, it is clear that the heritage of our evolutionary past has laid the ground for our present and future, and that the calorie-dense, predator- and famine-free world in which we live conspires to mean that many of us will struggle constantly to keep to a healthy weight. The idea that we can delve in to our pre-agricultural roots and devise a diet more in keeping with those simpler times in which we evolved is appealing and has generated a considerable industry to support it. As we have seen, the approach can work for weight loss, but it is a fiddly and expensive diet that likely provides no advantage over conventional calorie-controlled diets. It may also have some negative effect on our skeletons, but of course this could be reduced by taking supplements or seeking out specific nutrient-rich foods.

The crux of the paleo diet is that it is predicated on a single and simple fact: that we modern humans are identical to pre-agricultural humans. As an intervention and a philosophy, the paleo diet assumes that we have not evolved in the last 12,000 years, at least in terms of our ability to handle different foods. Given just how big an effect the development of agriculture has had on our environment, this seems on the face of it to be a pretty bold position to take. As we'll see in the next chapter, the development of agriculture and the availability of grains and dairy in fact led to considerable evolution and, later on, to yet more important mismatches.

CHAPTER THREE

Too Much Intolerance

The idea that we are perfectly evolved for a world that no longer exists, that somehow we have fashioned an environment for ourselves in which we are no longer fit for purpose, does stand up to some scrutiny when it comes to our growing waistlines. Although the slightly smug, 'I'm only fat because I'm so well adapted for famine' thrifty gene hypothesis lacks support globally, it does seem to explain one of the stand-out features of the global obesity crisis: the clustering of South Pacific Island nations in the top left-hand corner of the 'obesity–economy' graph. The alternative explanation, the drifty gene hypothesis, asserts that a lack of predation caused our upper weight limit to 'drift' upwards, unconstrained by strong selection pressures. Prior to the relatively recent environmental shift towards cheap, sugar-rich, fat-enhanced, calorie-dense food available on every street corner, this gradual drifting would have gone more or less unnoticed, although obesity was far from unheard of when people had sufficient access to plenty of food (Henry the Eighth being a prime example).

When we start trying to apply some very broad-brush evolutionary thinking to our diet, though, we start getting into trouble. The paleo diet approach 'works' in that a calorie-restricted diet will allow you to lose weight, but the diet assumes that humans have been stuck in an evolutionary rut for the past 12,000 years or so; that we are somehow the same as our Palaeolithic ancestors. As we have already seen in the South Pacific that is far from the truth. What is more, if the drifty gene hypothesis holds sway across the population, then the even greater release from predation that would have come in the past 12,000 years of settled human societies would hardly have reduced the drift (which is, remember, non-adaptive evolutionary change) in our upper weight limit. Evolution has caused other recent changes related to diet that influence more than just our fat metabolism. Our ability to physically process and digest different foods provides some of the very best examples of recent human evolution. These changes illustrate both how we can evolve relatively rapidly to fit in with environmental changes of our own making, and how very recent environmental changes have simply occurred too quickly for evolution to catch up. In summary, the development of agriculture provides us with a case study of how large-scale environmental changes lead to evolutionary mismatches and are sometimes solved by evolutionary responses, and sometimes not.

The rise of agriculture

Around 12,000 years ago something truly massive happened to the human environment. We finally worked out that growing plants is an awful lot easier than finding them and that keeping animals is a much better arrange-ment that hunting them down with spears and arrows. We didn't go from gathering berries and hunting deer to having a fully fledged farming system overnight, though. The transition from our hunter-gatherer phase to one when we were much more reliant on agriculture was relatively slow. It occurred in multiple locations and went hand in hand with other crucial developments like the widespread rise of urban centres and the beginnings of 'civilisation' as we know it. Of course, harvesting seasonal bounties from nature, like fruits and nuts, and hunting wild game didn't disappear from our food-gathering repertoire (and indeed are enjoying something of a resurgence thanks to the recent 'forager' movement) but for many, and especially those living in or near major population centres, there was no longer the need to source food directly from nature's larder full-time. With food production becoming increasingly centralised, we could start on our long but relatively rapid road to modernity. Make no mistake, the development of farming in the so-called Neolithic revolution was a very big deal indeed.

At that time in our history humans were already well dispersed. Africa, Europe, the Middle East and Asia were already populated and humans were well established through Indonesia and Australia. Around 20,000 years ago, the last so-called 'Glacial Maximum' ended and the ice sheets that covered great swathes of the Northern hemisphere prior to that time began retreating, never again to reach as far south as they had. As the ice sheets melted, humans headed northwards and populations also crossed the land bridge that connected what is now north-eastern Siberia with Alaska. Through this movement the peopling of the Americas began, and it marked the start of the period when *Homo sapiens* developed a truly global distribution. This widespread distribution inevitably meant that agriculture spread from a series of globally distinct centres, with the developing technology and knowledge spreading out from those centres to more far-flung populations over the course of several thousands of years. Attempts to account for the drivers behind the development of agriculture and its subsequent spread have resulted in a large number of competing, although not necessarily mutually exclusive, hypotheses. Undoubtedly, as we uncover more archaeological evidence and discover more about the climate and environment of the period during which agriculture developed, the details will become clearer.

Agriculture: really bad for our health

Growing our own food and developing the skills and knowledge to keep livestock is one of our greatest achievements, and without agriculture the modern world, and virtually anything we consider as a recent human achievement, would clearly be impossible. Agriculture serves primarily to free people from the daily necessity of finding food. What is more, if food production and distribution are reasonably well managed, and short- to long-term storage of food is possible, then food surpluses can be produced to even out the problems that might be caused by climatic issues such as the droughts and cold snaps discussed in Chapter 2. Taken together, potentially plentiful production and stable storage would seem to provide something of a paradise for a human population used to eking out a living from whatever could be found opportunistically in the environment. It therefore comes as something of a surprise to discover that the agricultural revolution went hand in hand with a dramatic decline in human health.[1]

Piecing together how we responded to the Neolithic agricultural revolution relies on integrating a wide range of evidence from different sources. Preserved teeth from Neolithic humans provide one such evidentiary source and, being relatively straightforward to interpret, the teeth tell quite a tale. Oral health was markedly poorer in humans subject to an early agricultural diet than their hunter-gatherer predecessors. Evidence can be found for reduction in tooth size, crowding of teeth, dental caries and increased occurrence of gum disease. These dental problems likely came about because of an important evolutionary change in human face shape that occurred during this period and that we met first in Chapter 1 when we discussed the characteristics of anatomically modern humans. One defining feature of modern humans is the development of a pronounced chin, which, combined with our relatively short jaws and steep foreheads, makes our proportionally smaller faces almost vertical. Palaeolithic, pre-agricultural humans had heavier, larger skulls than later Neolithic humans, but as we developed towards eating agriculturally derived food our masticatory needs changed. Archaeological evidence of grinding stones and cooking vessels during the Neolithic suggests that we physically processed grains and other foods, just as we do now, to render them more palatable and versatile. Further processing through more advanced Neolithic cooking than was undertaken in the Palaeolithic would have made foods even softer and easier to chew. By outsourcing much of our food processing from our teeth to our hands, Neolithic skulls no longer had to accommodate the larger masticatory (chewing) muscles required for arduous in-mouth processing. Over time this resulted in evolutionary changes to skull shape, proportions and angles. These changes included adjustments to jaw size that were not followed by changes in our dentition. With the same number of teeth cramming into smaller jaws attached to more graceful skulls, we ended up with dental overcrowding. Teeth that are crammed together create tight spaces in which bacteria can grow and dental caries can form. We entered a world of tooth decay.

There is another more subtle influence of processing on teeth wear that is apparent from archaeological material. 'Microwear' describes the tiny pits and scratches that form on the surfaces of our teeth and provide nooks and crannies in which the bacteria causing dental caries can reside. Processing food using stones introduces tiny, abrasive grit particles into our food that cause microwear of tooth enamel. When the sort of coarse foods eaten by hunter-gatherers are consumed regularly then microwear tends to be 'polished' out, but eating softer foods with grit prevents this polishing effect and can lead to caries.

Early agriculture didn't just lead to bad teeth. Hunter-gatherers had much smaller populations than early agricultural communities and a much wider range of dietary options, albeit a range closely constrained by the environment. Despite the vagaries of that environment, hunter-gatherers were actually less likely to suffer from diseases caused by nutritional deficiencies than those who were farming and reliant on a relatively narrow range of foods. Early adopters of agriculture depended on one to three crops, typically chosen from barley, wheat, millet, maize and rice. Cereals like wheat, barley and millet are great for carbohydrates, but they contain relatively few of the vitamins and minerals essential for adequate nutrition. They are, for example, low in iron and calcium, a problem made worse by the fact that they contain phytates, compounds that actually inhibit the absorption of iron, magnesium and, to a lesser extent, calcium from the diet. Maize is deficient in certain important amino acids and also inhibits iron uptake, while a diet that is heavy on rice can lead to vitamin A deficiencies. As well as being low in iron and calcium, in general these staples are low in protein. An obvious source of all these grain-deficient nutrients is meat, but we know that early agricultural communities ate significantly less meat than their predecessors. This has led some to suggest that early agriculture may also have led to zinc and vitamin B12 deficiencies. Iron and vitamin B12 deficiencies are primary causes of anaemia, and indeed signs of this disease can be seen in characteristic bone lesions apparent on some Neolithic skeletons.

Skeletons reveal other debilitating effects of the Neolithic revolution. We shrank, becoming shorter and less robust overall, with a stunting of growth apparent at a population level. In part, the lower physical demands of an agricultural lifestyle may have contributed to that reduction in stature, since our skeleton readily adapts to the prevailing conditions in which it must function; bone mass increases as we exert ourselves and put more stresses and strains on the different components of our skeleton. However, the 'lazy farmer' hypothesis doesn't explain many of the observations of Neolithic remains. Clear signs of incomplete enamel formation on teeth, evidence of 'growth arrest lines' on bones (visible lines that indicate periods when growth has ceased) and indications of weakened bones (osteopenia and in more serious cases osteoporosis) can all be related to malnutrition to some degree.

Dietary changes clearly caused problems in the early days of Neolithic agriculture, but developing farming also resulted in radical lifestyle shifts that went far beyond being generally less active than our hunter-gatherer predecessors. The development of larger communities and even cities acted to concentrate some human populations to create areas with both high population and high density. Humans are social creatures (more of which in later chapters) and we interact in complex ways. We work, play and live together, greeting each other, hugging, kissing, having sex, all the while speaking to each other with open mouths, spraying saliva as we go. The opportunity for these interactions increases greatly as our population density increases and with interaction comes the transmission of both ideas and diseases.

Smaller and more dispersed populations have far less opportunity for infectious diseases to spread and become epidemics than higher density populations. The towns and cities that developed as a consequence of agriculture put large numbers of people into very close proximity with each other and coupled with poor sanitation would have provided ideal conditions for disease outbreaks that would have been more or less unknown previously. Sure enough, analysis of the skeletons of Neolithic humans suggests that they experienced greater physiological stress from infectious diseases, the effects of which would have been exacerbated by poor nutrition.

Early mismatches

Early agriculture, depending as it did on a limited range of crops, led to a degree of malnutrition that had potentially serious effects, especially when combined with the social lifestyle changes that were also happening. These early farmers were very clearly 'unfit for purpose'. Brain power had allowed them to change their environment through a powerful combination of the realisation that crops could be grown and the ability to convert that line of thinking into a practical solution. We might think of the recent changes in our lifestyles as radical, and they undoubtedly are, but the shift to agriculture was arguably far more radical in terms of its overall impact. The major differences are that fewer people were affected (our population was perhaps only 5 million at the time) and that the shift happened across the global population over the course of a few thousand years. Nonetheless, the shift to agriculture provides a very useful example of how we can change our environment, render ourselves unfit for purpose and potentially evolve out of problems we make for ourselves. In many ways then, the Neolithic revolution could be a model for our current situation.

One nutritional problem strongly suggested by Neolithic bone evidence is a lack of calcium. Calcium is a metal that combines with phosphorus and oxygen to produce calcium phosphate. It is calcium phosphate that provides the mineral component of our bones and gives our skeleton strength. Our skeleton is a dynamic system with bone being laid down and resorbed all the time, and diets low in calcium can adversely affect the development and ongoing strength of the skeleton. Calcium gets into our bones via our diet, and patterns in modern-day dietary calcium intake vary greatly across the globe, as a study in the scientific journal *Osteoporosis International* has shown.[2] Northern European countries like the UK, Ireland, Germany and France, and countries that have recently developed through an influx of migrants from these countries, like the USA and Australia, have relatively high dietary calcium. Meanwhile, countries in South, South-east and East Asia, including countries with very high populations such as China, India, Indonesia and Vietnam, have a notably low calcium intake. Below 400mg of calcium a day is a known risk factor for developing osteoporosis and in this study China, Indonesia and Vietnam were comfortably below that level (an average of 338mg a day, 342mg a day and 345mg a day) and India only just above it (429mg a day).[3]

The global pattern of calcium intake provides us with a clue as to how some of our ancestors solved the Neolithic calcium crisis, because in countries where people have a diet rich in calcium most of it comes from dairy. In a study in the US, for example, an estimated 72 per cent of calcium comes either from 'straight' dairy (milk, cheese and yoghurt) or from foods to which dairy products have been added (including pizza, lasagne and other 'cheese on top' meals, and dairy-based desserts). The remaining calcium comes from vegetables (7 per cent); grains (5 per cent); legumes (4 per cent); fruit (3 per cent); meat, poultry and fish (3 per cent); eggs (2 per cent); and the less helpful category, miscellaneous foods (3 per cent). In the UK, the figure for straight dairy products is slightly lower (50–60 per cent) but we still get the bulk of our calcium from dairy.

The dominance of dairy in our diet is such that for those us who grew up in the UK, 'drinking milk' is pretty much synonymous with 'healthy bones'. When I started primary school in Devon in the late 1970s we had mid-morning 'milk time', when children were more or less force-fed their daily bottle of milk because it was 'good for bones'. Free milk for schoolchildren had initially come about because of the 1946 Free Milk Act, passed in response to earlier research that had identified the link between low income, malnutrition and under-achievement in schools. Milk for secondary school pupils was stopped for budgetary reasons in 1968 and the same financial pressure removed milk for children over seven in 1971. The Education Secretary at the time was Margaret Thatcher, a fact recognised later by the popular and undeniably catchy anti-Thatcherite chant 'Thatcher Thatcher Milk Snatcher'. In fact, documents from the time suggest Thatcher may have tried to save free milk, and was overruled by Prime Minister Edward Heath, but regardless of the details the 'free milk for children' debate did not end in 1971. Rumbling on through the 1980s, free milk for children lived on via various subsidies and Acts like the 1980 Education Act. Children, including my own, now benefit from free milk when they are under five years old through the Nursery Milk Scheme and subsidised milk thereafter through schemes like Cool Milk. The ingrained mantra, that milk is good for you, means that many parents pay for their children to drink milk at school as well as at home.

It is little surprise that milk has become revered as a healthy drink. For one thing, milk and dairy generally are wonderful sources of calcium. To get the same amount of calcium as you get in a 250ml (9fl oz) glass of milk (322mg) you would need to eat 685g (1½lb) of broccoli, which is at least two whole heads and maybe three if you like to trim off much of the thick stalk. Other non-dairy relatively calcium-rich foods include nuts (about 275g, 10oz to get the same calcium as a glass of milk) and eggs (12, give or take), so no one is denying that you *can* get calcium through non-dairy sources, but dairy is certainly a super-concentrated and reliable source. It is the consumption of large amounts of dairy that accounts for the high calcium intake in countries at the top of calcium league. We can also flip that fact around and in so doing account for low calcium intake in countries at the bottom of the league.

Milk and dairy do not feature much at all in the Asian diet and this absence largely explains the low dietary calcium intake observed in those countries. This situation is changing slowly. China in particular is greatly increasing its intake of dairy, especially so in urban areas, but it still lags far behind European countries.[4] That said, the shifting balance of the Chinese diet towards higher dairy intake is happening to such an extent that is becoming a concern to many environmental commentators because satisfying the demand could have negative impacts on climate change.[5] We are still unsure of the wider-scale health risks of a lower calcium diet in these countries because we lack meaningful large-scale studies of bone health in many countries. Where we do have data though, the implications are worryingly clear. A recent study of osteoporosis in China concluded that more than one-third of people over 50 are affected (34.7 per cent), compared with just 6.8 per cent of men and 21.8 per cent of women in the UK and broadly similar figures in France, USA, Germany, Spain, Italy and Sweden.[6] While correlation does not imply causation, it is extremely hard to ignore the dairy–calcium–osteoporosis link.

Milk saves the day?
The initial problems of the Neolithic post-agricultural diet were in part solved by improving agricultural ambition. Cultivating a wider variety of crops provided a better range of nutrients, while establishing trade links between different agricultural communities both near and far provided produce (and technology and knowledge) that allowed a more varied and balanced diet to develop. These different improvements took place slowly and incrementally around the world, but around 7,500 years ago a far bigger and more rapid development occurred. Around that time, most likely in central Europe, we started drinking milk or, more accurately, adult humans started drinking the milk of other mammals. Understanding when, how, why and where this dietary shift occurred shows us both how we can evolve to fit a changing environment and why our recently changed environment is not working out for a great many people across the world.

Drinking milk comes naturally to infant mammals, who instinctively root around to find nipples, teats or udders. Indeed, as any parent knows, infant humans will readily attempt to suckle on fingers, ears and noses. Most mammalian infants can be persuaded to suckle milk from a bottle if a suitable teat is provided. I once had the honour of bottle-feeding a beautiful suckling zebra called Mbezi. She greedily took two litres of horse milk (made from powder) in a good deal less than a minute from a soft-drink bottle with a length of rubber gas hose as a substitute zebra udder. The clear lesson is that infant mammals love milk, and so they should. The milk received from the mammalian mother is a highly nutritious food, and its rich mix of proteins, fats and sugar in the form of lactose supports very rapid growth in species as varied as the pygmy shrew and the blue whale. Where things get weird is the drinking of milk as an adult. You simply don't find this habit in non-human mammals, and the reason for the absence is that most adult mammals cannot properly digest milk. It is, quite literally, baby food.

To digest milk, and specifically lactose (the sugar in milk), requires an enzyme called lactase. Since the only function of lactase is to digest milk, and mammals only drink milk when they are suckling, the activity of that enzyme drops off massively after young are weaned. Evolution tends towards efficiency and if something isn't needed then it doesn't tend to be produced. To drink milk as an adult requires the adaptive evolution of what is termed *lactase persistence*, where lactase activity continues past weaning and into adulthood. Without that evolutionary change, in fact a genetic mutation, drinking milk as an adult will make you ill. This would have been the case in humans during the Neolithic revolution. Although there is evidence that we started to domesticate sheep around 11,000 years ago,[7] and goats and cattle only slightly later, we would not have been able to drink their milk, at least not in raw form or in any quantity.

Lactase persistence

It took around 3,500 years for our genetics to catch up with our environment and provide at least some of us

with the biochemical tools we needed to digest the
wonderful source of calcium (and fat and protein) that
we could squeeze out of our developing livestock herds.
In other words, although it took a while we were able
to evolve our way out of at least some of the problems
agriculture had created. This evolutionary solution
wasn't hit upon in all Neolithic populations though,
and even now lactase persistence is very far from evenly
spread throughout the world. In north-west Europe for
example, lactase persistence is high at 89–96 per cent[8]
but this declines gradually as we move south and east
through Europe. Only 17 per cent of Greeks, for
example, are predicted to be lactase persistent in a
study that combined data from well-studied populations
with a theoretical approach that 'filled in the gaps' for
those populations that had been studied less. The same
study also predicted 100 per cent lactase persistence
in Ireland.[9] Relatively high frequencies of lactase
persistence are found in some other populations,
including sub-Saharan Africa and the Middle East. In
Africa, the distribution of lactase persistence is patchy
and can be highly variable even between neighbouring
populations. For example, a study of nomadic pastoralists
(the Beja) living between the Nile and the Red Sea in
the Republic of the Sudan, and the neighbouring semi-
nomadic cattle-breeding Nilotic peoples in the south
of the country, revealed more than 80 per cent of Beja
were able to absorb lactose compared to less than 25
per cent of Nilotic peoples.[10] In Asia, the pattern is
clearer; the overall frequency of lactase persistence is
low and more or less uniformly so. In most Chinese
populations lactose persistence is less than 5 per cent

and a similar frequency is found in East Asian and Native American populations.

It can come as a surprise to many people who have grown up in cultures and nations that revere milk as a healthy food full of calcium, but most adults in the world simply don't have lactase persistence. Most people in the world cannot drink milk, at least not in the quantities that Europeans tend to. Of course, the pattern we see in lactase persistence in present-day populations pretty much mirrors the pattern we saw in the consumption of dairy around the world. We don't tend to eat foods that we can't digest and that make us ill. The pattern can also be used to help explain the story of how and why lactase persistence evolved, and why that relatively recent evolution has left modern populations both fit and in some cases unfit for purpose in the modern environment.

In Europe, lactase persistence is explained by a single mutation that has increased greatly in frequency over the last 7,500 years or so. Genetic studies show that around that time, in a region between the Balkans and Central Europe, the lactase persistence mutation began to increase and spread.[11] Since being able to digest lactose (and therefore consume milk) is only an advantage when there is a source of lactose available, it is likely that lactase persistence first began in dairying communities where the ability to consume milk would have provided a strong advantage. Consequently, we can also infer that cultural evolution of dairying as a farming practice co-evolved with the ability to drink milk; genes and culture co-evolved, with each encouraging the subsequent evolution of the other.[8] Communities that could drink milk would likely become more dependent on dairy, and their farming culture would then further change to accommodate this dietary preference. The presence of more milk in the environment and the advantages that it provided would lead to a selective advantage to the lactase persistence gene.[12] Fast forward a few millennia and we see some adult populations where regular consumption of large quantities of cow, sheep and goat baby-food is the norm.

In Africa and Asia, lactase persistence is both far rarer and more complex. Four known mutations are associated with it, and there are likely to be more. These mutations occur at different frequencies in different populations. The reasons for these population differences could be linked to the type and strength of the selection pressures for lactase persistence in different populations. In Europe it has been proposed that selection for milk drinking might have been linked to the fact that the lactose and vitamin D in milk enhance the absorption of calcium (from both the milk and other dietary sources). This might have been especially beneficial in regions of low light like Northern Europe because of the possibility of vitamin D deficiency. Vitamin D is essential for calcium uptake, and it can be synthesised in the skin from cholesterol in a chemical reaction that requires exposure to UVB radiation from the sun. Consuming dairy, with its high concentrations of both calcium and vitamin D, could have helped to compensate for lower levels of vitamin D from sunlight in gloomy Scandinavia, for example.[11] As appealing as this one-stop-shop 'calcium assimilation hypothesis' is, it is not supported by a study of the spread of lactase persistence through Europe. This study concluded that the need for a source of dietary vitamin D was simply not necessary to account for lactase persistence in Europe. Instead, the reason for there being a single mutation in this case is probably because of human history. Milk and products made from it are a nutritious source of calcium, and following the evolution of lactase persistence in the Balkans, there was a wave of population expansion across Europe from that region. This, at least according to some researchers, explains why across Europe lactase persistence is a consequence of the same single mutation. In contrast, in Africa and Asia a number of different selection pressures have been proposed for lactase persistence, including milk drinking as a source of hydration, as a simple source of calories and nutrition, and even as a potential mechanism to reduce the effects of malaria.[13] The diversity in lactase persistence in these populations is possibly also the consequence of their varied farming practices that did not rely so much on dairy. This differed from Europe, where dairy farming expanded more widely over the whole population.

Working out the precise details of the story of lactase persistence evolution is very much a scientific work in progress, combining genetics with knowledge of early human populations, their culture and migration. Regardless of the details, the evolution of lactase persistence would clearly have required a fortunate set of circumstances bringing together the latent ability of the environment to produce milk from animals kept primarily for meat, with the appearance of a mutation that allows adults to drink milk. Without that interaction between cultural farming practices and genetics, it is much less likely to occur. In China for example, where overall lactase persistence is rare, the historical tendency towards dairy farming and milk consumption is also rare.

For whatever reason, that gene–culture coupling never got off the ground.

The evolution of lactase persistence shows us very clearly that we can and have evolved in response to the changes in our environment that we have ourselves brought about. However, it also teaches us that these evolutionary solutions occur relatively slowly, across timescales of thousands of years. The evolution of fixes for mismatches is subject to the vagaries and differences of both our own imposed environment and the wider environment. The fragmented nature of human populations and the chance appearance of the necessary mutations mean that these changes have tended to occur unevenly across the globe. Even today, despite globalisation, the relative ease of long-distance travel and a gradual increase in wealth and economic freedom, 'humans' are not a single population. We are still fragmented to a large degree into different populations, and the environment and selection pressures differ greatly between, and even within, those populations. Any evolutionary changes that might occur today to get us out of trouble will still be influenced by the same basic forces and processes as in the past. The evolution of lactase persistence gives weight to the idea that we *could* evolve our way out of self-imposed problems, but it gives us really no hope at all that we *will*, at least in any useful timeframe.

Globalisation and the dairy problem

In the past, the fact that some strange people in far-off lands consumed milk would be little more than an amusing piece of trivia with which to impress your neighbours; an early version of the 'crazy foreigner' trope. The development of communications and transport technology over the past 50 years or so, however, has given a great many of us a far more sophisticated awareness of the rest of the world than was ever possible before. These technological developments have moved in step with the development of advanced medical science, the general push towards epidemiological medicine in developed nations and the widely acknowledged and often (though not always) evidenced link between health, nutrition and diet. That these advances have very often occurred in nations that also have lactase persistence means that the firm identification of milk as a healthy food, the association of milk consumption with a Western lifestyle and the global dissemination of the 'milk is good' message were inevitable.

The modern world allows a global reach for the message that 'milk is good', but this reach is a clear mismatch with the global pattern of the evolution of lactase persistence. This mismatch occurs at the level of nations and even entire continents but can also occur within nations with a high frequency of lactase persistence. The fact remains that across the world around two-thirds of people cannot consume raw milk, and even within Northern European populations the ability is not ubiquitous.

The recent rise of dairy in China provides a useful case study to show how very recent self-imposed changes in our environment can be mismatched with our evolutionary past. We already know that China consumes far less dairy than European nations or the USA, but from a very low level China's dairy consumption has been steadily increasing.[14] The UN Food and Agriculture Organisation estimated that China's consumption of milk increased from 26 kilocalories per person per day in 2002 to 43 kilocalories in 2005, which although still very low in comparison to Western nations does represent a near-doubling. The reason for this rise, which has only increased further over the past 10–15 years, is difficult to pin down precisely, although political pressure has certainly played a part. When in 2007, Chinese premier Wen Jiabao said, 'I have a dream to provide every Chinese person, especially children, sufficient milk each day', we can be sure that he wasn't talking to himself. Xuē Xīnrán, a British-Chinese journalist and author who writes under the pen-name Xinran, explored the rise of milk-drinking in her columns in the UK newspaper the *Guardian*. Published as a book in 2006, *What the Chinese Don't Eat* explains the 'dairification' of China as an aspirant phenomenon. 'Until China opened up, Chinese people had no idea about international standards. This is why people in the 1980s believed McDonald's was the best Western food,' she says. 'They believe that Westerners had a better life based on meat and milk.' That foreign lifestyles can influence domestic choices is undeniable, but whether this is a factor in the rise of China as a fledgling milk-drinking nation is debatable. Professor James Watson of Harvard University is an anthropologist who specialises in food-eating and China. He dismisses the idea that it is admiration for the West that has driven the rise in milk consumption, instead proposing that simple availability has been the key. In the past, because so few people could drink milk, no one produced milk. A nation that lacks lactase persistence is a nation than lacks dairy. The modern world has put dairy within most people's reach and this drives demand, at least according to Watson. He sums up this idea rather neatly as 'it [the rise in milk consumption] doesn't indicate they are becoming more Western, it just means they like ice cream.' But with all we now know about milk and lactase, they really shouldn't like ice cream, and neither should most people across the world. As wonderful as milk is for many, two-thirds of people just don't have the genetic tools to digest it. We tend to term such people 'lactose intolerant', although given they are in the majority it might be better to call them normal.

Intolerance

Without the right genetic tools in place to handle lactose as an adult, consuming milk and dairy will make you ill. The symptoms of lactose intolerance are not pleasant and include diarrhoea, stomach cramps, abdominal pain, bloating, nausea and flatulence. The triggering of these symptoms tends to be dose-dependent, with some people reacting poorly to very small quantities of dairy, like some milk in a cup of coffee, while others are able to tolerate much more before their body's reaction to lactose becomes a major problem. Not all dairy is equal, and processed dairy products like butter, yoghurt and cheese contain lower levels of lactose than raw milk, sometimes far lower. The process of making butter typically removes a great deal of lactose, and the fermentation processes involved in creating yoghurt and cheese also serve to reduce lactose, in some cases to less than 10 per cent of that found in the equivalent volume of 'raw' milk. This reduction in lactose, coupled with the dose-dependent nature of the lactose reaction, means that processed dairy can be consumed more easily by many people, and it is a rise in the consumption of these products and especially cheese that is a major factor in the rise of dairy globally.

Cheese contains calcium, which is a very good thing, but it is also a high-fat, calorie-dense food that is all too easy to add to a great many dishes. As sure as night follows day, the rise in dairy consumption has been linked by some to a rise in obesity in China, although other factors including the more widespread availability of Western-style fast food have also been implicated. If dairy is to blame for a rise in obesity then, with a knowing nod to the previous chapter, the story provides a nice example of how the complexity of the modern environment playing against our evolutionary history can cause us problems. Technology creates a world where the message that 'milk is good' spreads to populations that have not evolved to digest it; as a work-around, processed milk is consumed in quantities great enough to cause obesity, which is a far greater risk to health than low calcium levels.

The concept of processing milk to make it more palatable to those without lactose persistence is far from new. Archaeology combines with chemistry in unravelling the history of cheese-making, with analysis of carbon and nitrogen isotope ratios (see Chapter 2) from fatty residues on pottery fragments allowing scientists to determine whether those residues are from meat, fish, milk or fermented milk products. This technique was applied to pottery fragments from the Dalmatian coast of Croatia in 2018 and pushed back the earliest known cheese-making in the Mediterranean region to 7,200 years ago. It is likely that we will continue to push this date back even earlier with more discoveries and better analytical techniques.[15] The jump to making cheese from milk was presumably accidental at first (and hats off to those early pioneers, bravely tucking into the first cheeses) but would have given early dairy farmers two key advantages. First, cheese has a far better shelf life than raw milk, and second, cheese could be consumed by adults. The ability of adults to eat cheese would have driven an increase in dairying and, as we have already seen, this cultural change would have gone hand in hand with the evolution of lactase persistence.

Lactase persistence, and the flip-side of the coin, lactose intolerance, show how evolution has both prepared some humans and failed to prepare others for the modern world of near-ubiquitous dairy availability. It remains to be seen how the rise of dairy in lactose-intolerant nations will influence overall health, but it seems hard to argue that a rise in dietary calcium won't reduce osteoporosis or that increasing cheese consumption could lead to obesity. Whether either will have any long-term effect on human evolution, though, is questionable. Osteoporosis predominantly affects older people who have already passed on their genes before the problems hit. Evolution is largely blind to events that occur after reproduction ends, although having healthy and active grandparents could provide an evolutionary advantage, especially in the modern world. People with active parents who can help with childcare might be tempted to have more children than they would otherwise. If those grandparents are healthy and active because they have strong bones as a consequence of being lactase persistent then at least some of their children and grandchildren will likely also be genetically lactase persistent. More children and more grandchildren means more of an increase in the frequency of lactase persistence.

The modern world is seemingly providing an environment favouring healthy and active grandparents. We are seeing increases both in 'grandparental care', where grandparents assume primary custodial care of their grandchildren, and of grandparents acting as part-time child minders.[16] Full custodial care of grandchildren is driven by a range of complex societal factors but some of these, including addiction (Chapter 8) and economic uncertainties, can be clearly linked to the recent overall environment we have created. The 'grandparents as childminders' scenario is primarily driven by more straightforward changes in our recent economic and social environment. Higher house prices and other cost-of-living pressures over the last few decades often mean that two-parent households require both parents to work. Furthermore, social changes have taken us away from the 'traditional' household of the first half of the twentieth century and earlier, when mothers would be expected either not to have a career or to give up that career for child care. These changes in our environment exert all kinds of everyday financial and social pressures on us, but perhaps they are also exerting a selection pressure for grandparents with strong bones able to deal with the rigours of late-life child care.

As well as the increase in dairy farming, the agricultural revolution saw a rise in the availability and consumption of another component of our diet that is linked in the modern world to 'intolerance'. The ability to process the grain (in fact the seeds) of wild grasses into early flours by grinding and the realisation that grains can be sowed, collected and re-sowed were really all it took for cereal cultivation to develop. Once it had developed, perhaps 10,000 years ago in the Fertile Crescent of the Middle East (encompassing the modern-day countries of Iraq, Israel, Palestine, Syria, Lebanon, Egypt and part of Turkey and Iran), wheat cultivation quickly began to spread. With that spread came developments in farming technology that we still recognise today. Early farmers would have selected for better varieties both accidentally and deliberately, producing domesticated cereals with traits like increased grain production. Crop rotation, whereby different crops are grown in a sequence to limit soil nutrition depletion was developed, and leaving land fallow to allow recovery is even mentioned as a rule in the Book of Leviticus.

Cereal grains, especially whole grains (as opposed to highly processed and refined products like white flour) can be a reasonable source of protein and fibre, B vitamins, antioxidants and minerals like zinc and magnesium, although as we have already seen, early adopters of agriculture suffered from symptoms consistent with a lack of some of these nutrients. As well as providing many important dietary components in one hit, a diet rich in wholegrain cereals has been firmly linked to a number of important positive benefits that include a reduction in type 2 diabetes[17] and protection against various cancers.[18] However, increasingly in the modern world we seem to be revealing a major problem with cereals: a protein called gluten. Gluten comes from the Latin word for glue and it features heavily in the endosperm, the main component, of seeds like wheat, barley, rye and oats. The glutens (for there are many related proteins in the group) have a unique set of properties that give bread dough the combination of adhesion, holding itself together, and elasticity, allowing dough to rise.

Although the majority of us can consume gluten with no issues whatsoever, in some people gluten can trigger a whole range of symptoms and conditions that are grouped under the umbrella term gluten-related disorders (GRD). These include coeliac disease, non-coeliac gluten sensitivity, gluten ataxia, dermatitis herpetiformis and wheat allergy. Despite their vagueness and the fact that their use is discouraged medically,[19] the terms 'gluten intolerance' and 'gluten sensitivity' are still commonly used to refer to the set of symptoms brought on by consuming gluten. These symptoms can include diarrhoea, abdominal pain, bloating and nausea; indeed, many of the symptoms we have met already with lactose intolerance. Lactase persistence can mostly be linked to a single gene and a reasonably straightforward evolutionary scenario. GRD evolution is similar in some respects but more complex in others, although its rise can once again be linked to recent environmental changes of our own making.

Gluten intolerance and our immune system

Coeliac disease is an autoimmune disorder, with the immune system 'overreacting' when gluten is ingested. This abnormal response primarily causes inflammation in the small intestine, leading to that distinct and unpleasant suite of symptoms. Over time, this inflammation eventually causes damage to the lining of the small intestine and poor absorption of nutrients. In childhood, poor absorption caused by inflammation of the bowel lining can result in problems with growth and development, which is another serious symptom of CD. In adults, poor absorption can lead to iron-deficiency anaemia and symptoms including bone or joint pain, fatigue, seizures and (of particular interest from an evolutionary perspective) infertility. Coeliac disease (CD) then is both a serious and a permanent condition that affects around 1 per cent of people globally, although its broad range of symptoms are indicative of many different diseases and this means that CD is easily missed. There is also strong evidence that CD has been increasing worldwide and that the increase is not just because we're getting ever better at detecting and diagnosing it (which we undoubtedly are).[20]

CD is a genetic disease and more than 95 per cent of people with it have one of two types of a specific protein involved with our immune response. HLA-DQ proteins are present on the outer surface of some cells and form part of the complex system of signalling that occurs between our cells and our immune system. These HLA-DQ proteins bind to other proteins (antigens) derived from 'invaders', such as a disease-causing bacterium, and present them to immune system T cells. It is a way that cells can flag that they are in trouble and attract attention from the immune system. Such signalling mechanisms also allow the immune system to learn to distinguish between 'self' and 'non-self'. HLA-DQ proteins then play a big part in assisting our immune system in attacking foreign invaders while tolerating our own cells. Sometimes, though, it can all go horribly wrong.

There are seven different variants of the HLA-DQ protein, numbered DQ2 and then DQ4 through to DQ9.

The proteins are coded for by different variants (called alleles) of the HLA-DQ gene, and 95 per cent of people with CD have either the DQ2 or the DQ8 form. If you have neither DQ2 nor DQ8 then you are very unlikely to have the disease.[21] In the gut, gluten is broken down into lengths of amino acids called peptides and the DQ2 and DQ8 forms of the protein in the gut bind much more tightly than other DQ variants to these peptides. This tight binding means that those people with DQ2 and DQ8 forms of the protein are far more likely to activate T cells and then activate their autoimmune system if gluten is present. It is important to note that while most people who have CD also have these DQ2 or DQ8 copies of the gene, not everyone who has those genes develops CD. This shows that the onset and persistence of CD symptoms is environmentally triggered by some factor related to exposure to gluten proteins.

The only current course of treatment for CD is a gluten-free diet. This lessens the symptoms and in time can promote healing of the intestine, although there is evidence that damage to the small intestine may well remain even after adoption of a gluten-free diet. Given that CD levels hover below or around 1 per cent of the population, and that not all CD sufferers have the same level of symptoms, it perhaps seems strange that the terms 'gluten-free' and GF have gained such prominence recently. Supermarkets have entire GF sections, restaurants promote GF menus and the popular press is stuffed full with 'going gluten free' articles. The reason behind that rise is not primarily an increase in CD, although it has become more readily diagnosed and has increased in frequency over in recent years. In fact, a different, and difficult, condition called non-coeliac gluten sensitivity (NCGS) is the reason why bread and pasta have become the latest dietary villains.

When we speak casually of 'gluten intolerance' it is almost certain that we mean NCGS. The most common of the gluten-related disorders, NCGS has an estimated prevalence as high as 13 per cent and symptoms are similar to CD.[22] The big difference between CD and NCGS is that there are no diagnostic biomarkers for NCGS; in other words, we can't run a test to identify genes. Diagnosis of NGCS rests on testing whether patients have CD, testing whether they have the much rarer condition of being allergic to wheat and then asking if the symptoms clear up when the patient stops eating gluten. If the answers are no, no and yes then bingo, they have NCGS. It was first discussed in the 1970s, but it is over the past decade that NCGS has really come to the fore.

The simple fact is that we still don't have a clear understanding of NCGS, and it has taken some time to be widely accepted by medical practitioners. Undoubtedly the absence of any biomarkers leading to a clinical test that can definitively identify it (as we can with CD) has been a big factor in the relative reluctance to diagnose NCGS until recently. Indeed, it has been dismissed as a 'fad' by some and it is still questioned as a clinical entity by others, although this is changing as we learn more about it and about related diseases. The symptoms of NCGS are also similar to those of another problematic and still poorly understood condition, irritable bowel syndrome (IBS). Patients can find themselves caught in what some have called a 'no man's land', with symptoms that are explained neither by CD nor by IBS. Exactly what causes NCGS is up for debate and it may well be that different forms of NCGS exist, triggered by different factors. Proteins other than gluten in cereals might be triggers in some sufferers, and carbohydrates, collectively termed FODMAPS (fermentable oligosaccharides, disaccharides, monosaccharides and polyols), have also been implicated.

It is interesting to note that the rise of the internet has created an environment where people are much more able to self-diagnose, and in no condition is this more apparent than NCGS. Self-diagnosing NCGS based on personal experience and then treating it by adopting a gluten-free diet is made even easier by the huge rise of popular articles on the topic. One study identified a ratio of 4,598:1 of Google citations versus PubMed citations (a major medical science literature database) on NCGS, and the concomitant rise in the availability of GF products facilitates and arguably encourages self-diagnosis.[23]

With lactase persistence, the pattern of occurrence around the world gave us insight into its evolution and some of the issues that it, or the lack of it, may cause in the modern world. GRDs are far more global in their distribution, and with an overall frequency of 0.9 per cent there are a large number of global studies that support the notion that CD is one of the most common lifelong disorders affecting humans worldwide. Despite such widespread occurrence, patterns in the prevalence of CD, the frequency of HLA-DQ2 and the consumption of wheat do vary geographically. This variation can give us evolutionary insight but it also throws up an evolutionary paradox.

The paradox derives from two correlations. First, there is a correlation between the consumption of wheat and the frequency of HLA-DQ2, such that in regions where we see a high consumption of wheat we see a higher frequency of the form of the gene that causes CD. Second, there is also a correlation between the frequency of CD-causing genes and the duration of wheat consumption that shows the history of CD is related to the spreading of wheat cultivation following the agricultural revolution in the Fertile Crescent. The Palaeolithic pre-agricultural diet would not have contained such a high proportion of gluten-containing cereals, and diseases like CD would have little chance to express themselves. What we find after the development of agriculture though, is the increase of a condition that would have been a clear disadvantage to those who had it. This seems especially the case when we remember that in the early days of the agricultural revolution we weren't exactly thriving on our new diet and any additional pressure, like being nearly crippled with bloating and abdominal pain from eating cereals, would hardly have made sufferers evolutionarily fitter.

There was no gluten-free aisle in the Neolithic supermarket and with an increasing reliance on grains since that time, it is therefore a paradox that we find CD in modern populations at all. It might be expected that CD would have been selected out early in our agricultural history and, if still present, that the longer a population has had agriculture (and therefore presumed selection against CD), the lower the prevalence of CD should be. In fact, and paradoxically, that is not the case. CD is just as common in the UK, where cereal cultivation started around 4,000 years ago, as it is in Turkey, which led the way in cereal cultivation a full 6,000 years earlier. The frequency of HLA-DQ2 is actually higher in Turkey and Iran (where there is high wheat consumption that has continued for a long time) than it is in Finland and Ireland, both are which are late adopters and relatively low consumers.[24] The fact that despite clear negative effects on health CD has not been selected out is known as the 'evolutionary paradox of CD'.[25] In a sense it is two paradoxes rolled into one, since not only do we need to account for the puzzling persistence of CD, but we must also account for the fact that CD is increasing and doing so in areas with a high level of wheat consumption.

Resolving the paradox
One mechanism that could account for the continued existence of CD and the fact that seemingly 'bad' genes can end up increasing in frequency is called antagonistic pleiotropy. This describes the situation where one gene controls for more than one trait, and where at least one of those traits is beneficial and at least one is detrimental. In the case of CD, HLA-DQ2 and HLA-DQ8 are part of a distinct gene family that are located in a chromosomal region packed full of genes associated with our immune system. Chromosomes are those 'wobbly X' structures that form from the molecules of DNA that make up our genome. We have 46 of them in 23 pairs (one from each pair coming from each parent) and most of the time they are unravelled and invisible, only ravelling to form those distinctive tightly packed structures at a particular time during the period when cells divide to form new

cells. It has been proposed that the close physical association on chromosomes of HLA genes with another gene family essential for immune-system function known as KIR (killer cell immunoglobulin-like receptors) might have led to these two gene families evolving together as an integrated system of genes.[26] Strong selection for a good immune system associated these genes together and in the absence of gluten the HLA-DQ2 and HLA-DQ8 variants work just fine. It is only when we went and invented cereal-based agriculture and loaded our gut full of gluten that these variants became a problem. By that point though, the advantages they gave to immune systems of the past had locked them into a 'if they're coming, then I'm coming along too' relationship with KIR genes. So, HLA-DQ genes overall give benefit by aiding in immunity against some pathogens, but some variants (HLA-DQ2 and HLA-DQ8) have a cost that is only expressed when we changed to a dietary environment dominated by gluten.

Another potential clue for the persistence of HLA-DQ2 (which as we've learnt accounts for 95 per cent of CD cases) is that the gene has been shown to provide some protection against dental caries, which you'll recall were a major health implication of early agriculture (Chapter 2). Dental caries result from a three-way interaction between the owner of the decaying teeth, their diet and the bacterial cultures they support on their teeth. The increase in carbohydrates in the post-agricultural diet, together with increased tooth crowding, led to a higher frequency of tooth decay, which eventually prevents individuals from being able to eat properly. The 'DQ2-dental caries protection' hypothesis proposes that people with the HLA-DQ2 mutation would have had some protection against dental caries and might therefore have had higher survival and higher reproduction than those who lacked this mutation. We still need to find out much more about this association and we do not yet know the mechanism (it may be linked to the clearance of 'sticky' gluten peptides from the mouth), but if it turns out to be correct then the mutation that causes CD may have initially been subjected to positive selection because of the harmful effects of the exact same gluten-rich diet that triggers CD.[27]

Antagonistic pleiotropy and positive selection for protection against dental caries can account for the persistence of CD, but they don't account for the second component of the paradox: the fact that CD is increasing in recent times even if we account for increased diagnosis. Evidence from a study in Sweden seems to point us towards what is now a familiar category of explanation for this phenomenon: very recent changes leading to a mismatch between our modern environment and that in which we evolved.

Sweden experienced a rise in CD in children under two years old between 1984 and 1996 known as the 'Swedish epidemic'. Rates increased three-fold, eventually reaching a rate higher than had been seen in any other country at that time. There was then a sudden drop, with CD in that age group returning to the baseline level of the early 1980s. Analysis of the epidemic revealed that the rise was associated with two factors. First, there was an increase at that time in the amount of gluten present in early weaning foods. Second, there was a shift in the pattern of exclusive breastfeeding and the timing of the introduction of gluten to the diet. The study showed that children who were still being breastfed at the time when gluten was introduced had a lower risk of developing CD.[28] The problem is that recent studies using larger populations and randomised trials showed no increased risk of CD in relation to gluten being introduced and no protective benefit of breast-feeding. With every step forward in our understanding we seem to take another step back and it is instructional that such a prevalent disease, which appears on the face of it to be rather simple, can be so complex. While we are gaining good ground in terms of understanding the disease itself, getting to grips with the environmental landscape in which it is triggered is still proving to be difficult.

Another factor that has been linked to CD is how you are born, specifically whether you were delivered by caesarean section or vaginally. Again, the situation is complex with some studies revealing a link between delivery modality and CD, and other studies finding no link. It may seem curious that the mode of delivery could be thought to influence the triggering of a disease, but the connection here is with the bacteria that dwell within you.

The colonisation of our gut with bacteria that have, of late, become known as 'friendly bacteria' happens as we grow and develop. Genetic factors and our environment, notably our diet, all have profound effects on our inner ecosystem, but both mode of delivery (caesarean section versus vaginal) and early feeding (breastfeeding versus formula) also have a role in determining which bacteria colonise our gut. As we are learning, our inner ecosystem has profound effects on our immune system. Given the clear links between that system and CD, it is obvious that we should be looking for links between the gut microbiome (as our bacteria are collectively known) and CD. You will not be surprised to learn that the situation is again complex, but the picture that is beginning to emerge is that CD is linked to a reduction in beneficial bacteria species and an increase in potentially harmful species.[29] The change that our recent environmental shifts have wrought on our inner ecosystem is something we shall return to in the next chapter.

The evolution of agriculture was a massive environmental change for Neolithic humans, and it led initially to a considerable mismatch between our evolutionary history and our dietary environment. Meanwhile, accommodating a radically different diet led to evolutionary changes whose influence is very much apparent today, when we consider our ability or otherwise to consume the two primary dietary changes that agriculture brought upon us: dairy and cereals. The ability of some of us to continue to digest lactose in adulthood has combined with technological, historical and social factors to produce a 'dairy rich environment' across the world that clashes with the evolutionary history of many and brings potential problems (lactose-induced illness and obesity) as well as benefits. Agriculture also made possible a gluten-rich environment that ironically may have initially selected for genes that now combine with aspects of our modern environment to cause serious problems for significant numbers of people. In both cases, our recent trend towards globalisation and the 'homogenisation' of human culture towards a Western-style diet combine with the high relative availability and desirability of certain foods to create what is quite literally a toxic environment for some. As well as showing how environmental shifts lead to mismatches with our evolutionary past, what these stories of dietary intolerance really highlight is just how diverse different groups of people across the planet still are, even in the modern world.

CHAPTER FOUR

The Shifting In-vironment

Changes in our environment have clearly affected us profoundly. Changes that occurred in the relatively distant past, for example related to the development of agriculture, have resulted in compensatory evolutionary changes but much more recent shifts outstrip any potential evolutionary response. One area of our biology where the negative effects of recent environmental change are becoming ever more noticeable, and our awareness of their significance ever greater, is not about the environment as a factor that is 'happening' to us, at least not directly. Rather, the environment in question *is* us.

Recent human-induced changes to our external environment, especially our diet, are influencing the environment inside our gut, where bacteria crucial to our survival and well-being dwell. Unpicking how our modern lifestyle is counteracting thousands, and even millions, of years of co-evolution with our microscopic passengers involves exploring something we have already touched upon in the previous chapter: our immune system. But before we can get to that, and how we have messed it up with our modern ways, we need to consider what bacteria are doing in our gut in the first place.

Bacteria, simply incredible: humans, simply habitat

Bacteria are incredible. Their cells may be much smaller than ours, and lack the fancy membranes, nuclei, mitochondria and other structures that our cells have, but they nonetheless get the job of life done very effectively. For one thing, they are biochemical marvels and thanks to some neat molecular pathways there are bacteria that can happily make a meal of crude oil, rubber and even plastic. Their adaptability is not just dietary. Their simplicity, combined with their flair for metabolic innovation, allows bacteria to exploit virtually any habitat. Whether it be rocks more than a kilometre (over half a mile) deep in the Earth's crust, the bottom of ocean trenches, hot springs or glacial ice, you are likely to find bacteria thriving where nothing else can. Not all bacterial habitats are quite as extreme though and organisms, including us, provide a much more amenable and easier habitat for them to colonise. As relatively large animals with a high and constant body temperature we are teeming with them.

As soon as we start thinking of ourselves as a habitat then we have to start thinking about ecology, the scientific study of how organisms interact with each other and their environment. Regardless of the scale at which we are operating, understanding some of the basic properties of the environment is crucial to understanding those ecological interactions and the diversity they can promote. Diversity in ecological systems is often driven by what can be thought of as habitat complexity. This is the wholly intuitive notion that a more complex environment offers more places to live and more means by which to make a living (more 'niches') than a simpler environment offers.

The niche, sometimes thought of as the 'address and profession' of an organism, is a central concept in our understanding of ecology and it is a concept we can apply just as well to our gut as to a tropical forest. Like a tropical forest, our gut is not a uniform environment. There are a number of different, discrete sub-habitats within the gut with different physical, chemical and biological conditions offering a wide variety of potential niches to those organisms able to cope. So, to our microbiome (as the assemblage of bacteria that live in and on us is collectively known) our bodies provide a structurally and chemically diverse range of habitats.

Even our seemingly smooth and uniform external surface, our skin, is far from homogenous and offers a wealth of different nooks, crannies and surfaces on and in which bacteria can live. Venturing inside, our mouth also offers a diverse range of habitats (teeth, the gaps between teeth and between teeth and gums, the gums themselves, the upper and lower tongue surfaces, the roof of the mouth, the back of the mouth, inside the lips and so on). Further inside, the contrasts between different internal habitats are even more striking.

The stomach, for example, is highly acidic. Muscles periodically contract to churn the contents of chewed food and hydrochloric acid laced with protein-digesting enzymes, creating a regular cycle of emptying and refilling that creates a rhythmic dynamic like a rock pool emptying and filling with the tide. From the stomach we move through into the duodenum, the first section of the small intestine. Around 30cm (12in) long, it is the smallest component of the three sections that make up the small intestine; the other sections (the jejunum and the ileum) extend to around 7m (23ft). The small intestine is all about chemical digestion (breaking down food into its molecular components) and absorption through the huge number of small finger-like projections called villi that line the walls. The small intestine is not a real stamping ground for bacteria, and overall there are fewer than 10,000 bacteria per millilitre here. That might sound like a lot but a gram of soil (a not dissimilar volume) might contain 40,000,000. If bacteria do start thriving in there, this can cause a condition known as small intestinal bacterial overgrowth (SIBO) resulting in, among other symptoms, nausea, constipation, diarrhoea, bloating, abdominal pain, excessive flatulence and steatorrhea, an unpleasant sticky form of diarrhoea caused by fats not being absorbed properly.

Moving into the colon (also sometimes called the large intestine), we find a drier and less nutrient-dense environment. Water is clawed back, creating stools that are pushed along through the action of muscles in the colon wall towards the rectum and then, eventually, exit the body through the anus. The colon is where the bulk of our microbiota hang out.

Outnumbered, but not as badly as you think

The gut is a lengthy and varied environment offering considerable space and opportunity for any bacteria able to take advantage, and a great many do. So considerable are the habitat opportunities that we offer, it is often said that bacterial cells in and on us outnumber our own cells by a factor of 10:1. This figure is both startling and memorable and it has been very widely repeated in books (including one of my own), papers, articles, TED talks and

TV shows. It is also wrong. The number came from a back-of-an-envelope calculation that has been traced back to a 1970 paper wherein, unsupported by evidence, an estimate of the number of microbes in a gram of intestinal contents was given as 100 billion. Combined with an estimate of 1,000g of gut contents, simple multiplication gives us 100 trillion gut microbes. Already, given our knowledge of the heterogeneity of our gut, this feels shaky. If the gut isn't uniform then might we expect bacteria populations to be uniform throughout? Almost certainly not. Nonetheless, seven years later in another paper this 100 trillion total was compared with 10 trillion cells in our own body, a number this time drawn from the pages of a textbook but again unsupported by evidence. Working with numbers in the trillions can be tough but dividing 100 trillion by 10 trillion is easy, and lo and behold we have the magical ratio of 10:1.

It wasn't until recently that some fact-checking was done on the numbers underpinning the calculation. When you think about it, determining the number of cells in the human body in any meaningful way is basically impossible without a certain number of important caveats. We clearly have to allow for a massive variation in size across humans and therefore a massive variation in cell count. Height and weight are obvious factors but males and females also differ and of course there is a huge variation imposed by age, which itself affects height and weight up to a point. In 2016, Ron Shender, Shai Fuchs and Ron Milo revised estimates both for the number of cells in the human body and for the number of bacteria. Using a 'reference man', an adult male weighing 70kg, they came up with a figure of 30 trillion cells, some three times the value previously assumed. In contrast, they downgraded the estimate of the number of bacteria from 100 trillion to 38 trillion. Overall, their estimates put the ratio at 1.3:1.[1] The fact that they also estimated the total mass of bacteria in our bodies at just 0.2kg (0.4lb) shows very effectively just how small bacterial cells are compared with our own.

The pursuit of microbial factoids and convenient numbers is ultimately a little pointless. As science writer Ed Yong pointed out in an article in *The Atlantic* responding to the 2016 paper, 'These new estimates might be the best we currently have, but the studies and figures … come with their own biases and uncertainties. My preference would be to avoid mentioning any ratio at all – you don't need it to convey the importance of the microbiome.'[2] I agree; when we are considering the role that the microbiome plays in our life we don't need a handy and surprising number to hammer home the point that without these bacteria we are dead.

What did bacteria ever do for us?
Although we provide a wonderful habitat for our gut bacteria, the relationship we have with them is far from a one-way street. First, they assist us greatly in digestion. The mouth, stomach and small intestine, with their triple digestive arsenal of chewing (physical breakdown), chemical degradation and enzyme action do a reasonable job of breaking down food into components that can be absorbed and used by our bodies, but they don't do a complete job.

We have very effective enzymes for breaking down proteins, but our ability to break down some carbohydrates, especially the complex branched sugars that are typically found in fruits and vegetables, is poor. We largely lack the molecular tools needed to deal with these molecules but some bacteria can digest them, breaking them down into molecules like glucose (widely used throughout our body), acetic and propionic acid (used by our liver and muscles) and butyric acid (used locally by cells lining the colon). These bacteria can also digest (and thereby recycle) the complex branched carbohydrates typically found in the mucus that we produce in considerable amounts from our gut wall to ease the passage of material through our intestine.

Our gut bacteria assist us in other ways. Vitamins are substances that are vital in very small amounts for our bodies to function, but we cannot make them. Our inability to make vitamins means that we have to get them from our food, and poor eating habits can lead to vitamin deficiencies and to diseases like scurvy (a lack of vitamin C) and rickets (a lack of vitamin D). However, there are certain species of our gut bacteria that are able to synthesise and subsequently supply to us vitamins such as folic acid (vitamin B_9, essential for making and repairing DNA, cell division and growth), biotin (vitamin B_7, required for the synthesis of a number of important molecules in the body), vitamin B_{12} (involved in DNA synthesis, fatty acid and protein metabolism and in the functioning of our nervous system) and vitamin K_2 (necessary for synthesising the proteins needed for blood coagulation). Bacteria also make it easier for us to absorb metals from our food. Our physiology and anatomy fundamentally rely on metals like calcium, which is found in our skeleton and essential for muscle and nerve function, magnesium, which is vital for energy metabolism, and the iron required for haemoglobin which transports oxygen in our blood. Fatty acids produced by bacteria digesting food in our gut make it easier for us to absorb these metals from our diet.

Our gut bacteria also assist us by suppressing the growth of harmful, pathogenic bacteria. Such bacteria cause harm by invading cells in our gut lining and, in some cases, subsequently invading other cells in our bodies. Beneficial species of bacteria stick to the lining of the intestine and in so doing they use up most of the available space and exclude other species. This produces a barrier effect, with potentially harmful invading species struggling to establish in the 'lawn' of beneficial species. Resident species, selected for their ability to thrive in the gut environment, are also far better at competing for nutrition in the gut. In a further twist, by fermenting complex carbohydrates into simpler molecules, resident species can produce substances (like lactic acid and fatty acids) that subtly change the environment within the gut to those that suit themselves and hinder competitors. As well as these largely passive actions, resident bacteria can be more assertive, producing toxins called bacteriocins that actively inhibit the growth of other bacteria.

Back to immunity school
Without our gut bacteria we can neither break down nor digest substantial amounts of the carbohydrates we ingest. They also provide a drip feed of some vitamins, increase our ability to absorb metals and prevent harmful bacteria from proliferating. In exchange we offer them a relatively safe home and ideal growing conditions.

However, for this relationship to thrive we need to make sure that they don't come under attack from our immune system. By achieving this harmony, we also need to make sure we don't leave ourselves vulnerable to attack from harmful bacteria.

Our immune system is a network of cells, tissues and organs that work together to attack and destroy invaders. White blood cells, or leukocytes, are one important part of the system. Made in bone marrow and the spleen, and stored there and in our lymph nodes, white blood cells circulate through blood vessels and the lymph system, constantly on the lookout for potentially problematic invaders.

There are two basic types of white blood cell: phago-cytes and lymphocytes. Phagocytes ingest invading cells. An important group of phagocytes are neutrophils, which are the most common phagocytes and the cells that target bacteria. If we have a bacterial infection, then the number of neutrophils increases in response to the increasing threat. The second type of cell are the lymphocytes and there are two types of these, B lymphocytes and T lymphocytes. Both cells are continually developing in our bone marrow. While B lymphocytes stay in the marrow, T lymphocytes move off to mature in either the thymus, a small organ located in front of the heart and behind the breastbone, or the tonsils. Their role is described below.

An invading bacterial cell is recognised as an invader because molecules on the outside of their cell membrane differ from the molecules our own cells have. These recognition factors are termed antigens, and antibodies are produced when these antigens are detected. Antibodies 'stick' to the invader's cell membrane and mark them out for subsequent killing by the phagocytes. This basic model of the immune system is sound and many bacteria are indeed killed this way, but there are other mechanisms whereby invading bacteria can be hunted down. For example, when some bacteria invade they can be targeted by specific immune proteins called complement proteins. These proteins can recognise antibodies and bind with them, causing further proteins to come and join the party, collectively becoming a membrane attack complex (MAC), a kind of small elite Special Forces protein team that can breach the cell membrane and eventually cause the invading cell to collapse.

Some of our immunity is present from birth. This innate immunity involves white blood cells sniffing out invaders and neutralising them. The innate immune system can recognise and deal with certain kinds of infection without any need to learn to distinguish 'good' from 'bad'. Any system of learning is time-consuming, and with bacteria able to multiply very rapidly it is essential to have a general strategy to deal with some problems quickly and effectively. This system doesn't work, though, if we are invaded by something that our white blood cells can't recognise. On the other hand the adaptive immune system can cope with novel invaders, although there is an inevitable lag period when we are first exposed. The adaptive system makes use of B and T lymphocytes. B lymphocytes can create antibodies that bind to these invaders and together with the T lymphocytes the cells can mount an attack. What is particularly useful with this system is that it remembers these novel attackers and if they are encountered again the response can be much faster, without the need to re-learn. Assuming we recover, we are unlikely to suffer from the harmful effects of those invaders (such as the virus that causes measles) more than once. The immune system 'remembers', and this is why immunisation programmes have proved to be so effective.

At first glance, the immune system might seem to be something of an obstacle for gut bacteria trying to set up home within us. It used to be thought that gut bacteria were essentially isolated from our immune system, only coming into contact with it when they breached the gut wall and entered the 'body' properly. However, microscopic examination of the intestines of mice, and subsequently humans, revealed that gut bacteria live in crypts in the intestinal wall that put them into very intimate association with the immune system. Research into how bacteria are able to reside in such a potentially hostile environment has revealed an elegant chain of molecular communication and responses. Bacteria produce complex sugar molecules called polysaccharides that are diverse in structure and have the potential to be recognised by other cells. Such molecules on the surface of bacteria are recognised by one of the cells of the adaptive immune system, the regulatory T cells, or Treg cells. Treg cells usually prevent the immune system from reacting to and attacking the body's own cells. If you have a problem with this system, the immune system can no longer distinguish friend from foe and can attack everything, causing an autoimmune response that can lead to autoimmune diseases.

With 'friendly' bacteria, the polysaccharide signal is detected by receptors on the surface of the Treg cells, and the cell is activated to suppress the activity of another type of T cell, T helper cells. The T helper cell is the assassin and the Treg cell is the guard. When the 'friend' bacterium appears at the door, the guard Treg cell recognises it and tells the assassin T helper cell to back down and leave it alone. Normally, triggering receptors on the surface of Treg cells activate the pathway that results in the elimination of bacteria. We have co-evolved with some bacteria, however, to produce ways that allow us to keep a tight security system active against harmful invaders while recognising types that are beneficial. Our adaptive immune system has to learn to identify these beneficial bacteria, but once it does they are free to stay, to the mutual benefit of both parties. So, at a healthy equilibrium your gut bacteria are vital for your immune system; essentially, they a *part* of your immune system. But it can go wrong and research is finding more and more links between a seemingly disconnected set of diseases (many of which seem to be on the rise) and gut bacteria. As with gluten-related disorders in Chapter 3, it is our immune system that is the link between gut bacteria and consequent symptoms and it is recent changes in our environment, and knock-on effects for our 'in-vironment', that is causing the increases we can observe.

When things go wrong
Unravelling the role that gut bacteria play in autoimmune diseases has been a major focus of biomedical science over the past decade or so. Thanks to these efforts, we have now reached a point where we can produce evidence of a surprising number of links between gut bacteria and

health, although it is notable that a recent review of the topic ended with a cautionary note: 'More robust, well designed clinical trials coupled with detailed mechanistic work will be needed to accurately intervene for the treatment and perhaps even prevention of autoimmune diseases.'[3] So, we are making headway but we are still very much at the start of this particular journey in human medicine.

Chief among the diseases in which a dysfunctional gut bacterial community is implicated is inflammatory bowel disease (IBD). IBD is an umbrella term that covers two conditions, ulcerative colitis (UC) and Crohn's disease. Crohn's disease is a long-term condition that causes inflammation of the lining of the digestive system. It can affect anywhere from the mouth to the anus but it most commonly occurs in the ileum (the main section of the small intestine) and the large intestine, sites that also, interestingly enough, harbour the highest abundance and diversity of bacteria. Crohn's disease causes a range of symptoms including diarrhoea, abdominal pain, blood and mucus in stools, weight loss and extreme fatigue. Symptoms can be mild or even disappear completely when the disease is in remission, only to flare up later and cause considerable problems. These symptoms can eventually cause damage to the gut that requires surgery to repair. Living with Crohn's is, by all accounts, miserable and it is increasingly common in the developed world. Ulcerative colitis has similar symptoms to Crohn's and this can cause problems with both diagnosis and treatment. While Crohn's can affect anywhere within the digestive tract and can damage all layers of the intestine, UC is focused on the colon and, more specifically, the top layer of the lining of the colon, although it can also affect the rectum. UC causes inflammation and, as the name suggests, ulcers develop on the lining of the colon. Crohn's can also cause ulcers but their distribution tends to be patchy, whereas in UC ulcerative damage appears in a more or less continuous pattern.

Although the conditions are similar, and related in many ways, there is a potentially important distinction between Crohn's disease and UC. UC is considered by many to be an autoimmune condition where the immune system goes rogue and, unable to distinguish friend from foe, attacks the cells lining the colon. Crohn's disease does not appear to be an autoimmune disease (so the immune system is not triggered by the cells of the body itself) but is instead regarded as an immune-related disease. You will note the cautionary use of 'considered by many' and 'does not appear', since both the diseases that comprise IBD remain relatively poorly understood. Perhaps the safest way to regard them currently is as 'immune-mediated inflammatory diseases', a catch-all if slightly vague term that has the advantages of being both accurate and descriptive.[4]

Determining the causes of IBD is still very much a work in progress and a single, simple explanation seems unlikely to develop, at least at the moment. That IBD might have complex causation is not perhaps surprising because the gut isn't a single simple system and is governed by at least three factors: our genetic makeup; our environment, which includes our diet, life history and lifestyle; and our gut bacteria. These factors act in concert in both a healthy and an unhealthy gut and a current favoured hypothesis for IBD combines all three, proposing that IBD occurs as an abnormal immune response to bacteria in the gut of genetically susceptible individuals that have been triggered in some way to become symptomatic.[4]

Evidence for the role of gut bacteria in IBD comes from a range of sources, including germ-free animals (laboratory animals, most commonly mice, raised in highly controlled conditions such that they have no bacteria in or on them) and human studies. Some of the more convincing evidence includes the fact that colitis in germ-free animals is significantly lowered, or absent altogether, and that a similar effect is found with antibiotic treatments that act to eradicate all gut bacteria. There is also some convincing correlative evidence from examining the composition of communities of bacteria, the overall assemblages of species that are present in the gut. In animal-based studies of IBD, certain species, notably *Bacteroides, Porphyromonas, Akkermansia muciniphila, Clostridium ramosum* and species belonging to a group called the Enterobacteriaceae (which includes *Escherichia coli*, perhaps the best known of the gut bacteria) are more predominant in cases of IBD and tend to be associated with higher levels of inflammation. Similarly, in human cases of IBD there may be increases in *Bacteroides* and Enterobacteriaceae and a decrease in Firmicutes (a group of bacteria that includes *Lactobacillus* and *Clostridium*) in comparison with people unaffected by IBD. There is also an overall trend towards lower diversity (fewer species present) in those with IBD than those without.

There are some further gut bacterial trends that link to IBD and strengthen the evidence base for the gut bacteria–IBD link. Some studies show a reduction of certain bacteria in IBD that produce short-chain fatty acids (SCFA) by digesting the complex carbohydrates found in the plant material we eat. These SCFA afford some protection to the epithelial cells that line the gut, inhibiting inflammation (both good things in preventing the symptoms of IBD).[5] These species include *Faecali-bacterium prausnitzii*, *Odoribacter splanchnicus*, *Phascolarc-tobacterium* and *Roseburia*. *Faecalibacterium prausnitzii* and *Roseburia hominis* also both produce an SCFA called butyrate, which is known to induce the formation of regulatory T (Treg) cells that you will recall from earlier usually prevent the immune system from reacting to and attacking the body's own cells. A high presence of both of these species is related to a low level of UC, and vice versa. It is also interesting to note that a reduction in *F. prausnitzii* has been linked to the recurrence of Crohn's disease occurring after operations (where antibiotics wreak havoc with our gut bacterial community), and adding the species to mice reduces gut inflammation. These, and many other studies, contribute to a growing evidence base that is underpinning an increasingly convincing narrative that gut bacteria, and the communities they form, are intimately linked with IBD and that it is the bacterial–immune system connection that is of prime importance.

The modern mismatch

It is the alteration of bacterial communities and the disturbance of the bacterial–immune system link that we must account for if we are to explain the rise of IBD as a mismatch caused by recent environmental changes. As with previous examples, an examination of the global pattern of the disease, and the changes that have occurred over recent time, is valuable in piecing together the story. In the case of IBD the patterns are relatively simple and clear. Over the last century, IBD has increased in Western countries but has now more or less reached a plateau in North America and Europe at around 0.3 per cent, with much of that rise occurring since the 1950s. The rest of the world lagged some 50 years or so behind this rise, but countries in the Middle East, South America and Asia have seen a recent rise in the incidence of IBD, making it a global disease firmly linked to newly industrialised countries becoming more Westernised.

Recent environmental change, broadly bracketed as 'industrialisation', mediated through the tightly co-evolved gut bacterial–immune system connection, is being held to blame for the rise in IBD. Given that IBD is intrinsically gut-related it makes perfect sense that the bacterial inhabitants of the gut might have a role to play in it. However, there are other immune-mediated inflammatory diseases in which gut bacteria are implicated. Patients with new onset rheumatoid arthritis, an autoimmune disease that attacks the cells that line joints, causing inflammation and pain, have gut bacterial communities enriched with *Prevotella copri* at the expense of *Bacteroides* species. Changes in gut bacterial communities, with enrichment of some species and diminishment of others, has also been seen in studies of patients with ankylosing spondylitis, an inflammatory condition affecting the spine. Recent work has highlighted connections between gut bacteria and multiple sclerosis. The cause of multiple sclerosis remains unknown but it is considered to be an autoimmune disease, with the immune system acting to destroy the fatty substance (myelin) that sheaths nerve cells in the brain and spinal cord.[7] Gut bacterial community imbalance has been observed in patients with multiple sclerosis, and the connection may be mediated by an enzyme common to us and to our bacterial passengers. T cells of our immune system react to an enzyme called GDP–L-fucose synthase, which is formed in human cells and in the cells of certain gut bacteria found in patients suffering from multiple sclerosis.[8] The evidence suggests that T cells are activated by this protein in the intestine. From there the cells make their way to the brain, where they come across the human version of their target antigen (the enzyme), causing inflammation and the subsequent symptoms of MS.

Our understanding of the influence of our on-board microbial communities on our health is increasing, not just in terms of the detail and mechanism underpinning that influence, but also in terms of the range and scope. For example, the term 'gut feeling' has a greater resonance as evidence accrues for the relationship between our gut bacteria and our mental health. Studies in mice have shown that gut bacteria can influence behaviour and small-scale studies of people with depression has indicated that the disease could be linked to changes in gut bacterial communities, just as gut community structure is related to inflammatory diseases mediated by the immune system. A larger-scale study was able to take advantage of a cohort of more than 1,000 Belgians who had initially been recruited in an attempt to quantify a 'normal' gut community. Some people in the group (173 of the 1,054) had been diagnosed with depression or had scored poorly on a quality of life survey and researchers were able to compare their gut bacteria with others in the cohort. They found that *Faecalibacterium* and *Coprococcus* bacteria were consistently associated with higher quality of life indicators. What is interesting is that both these bacteria are butyrate producers, a short-chain fatty acid we have met already when considering IBD. Butyrate, if you remember, encourages the formation of regulatory T (Treg) cells that prevent the immune system from reacting to and attacking the body's own cells.[9] On the other hand, *Coprococcus* and a species called *Dialister* were depleted in people with depression. These effects were evident even after correcting for the confounding effects of factors like age, sex and the use of clinical antidepressants.[10] Another notable finding was the positive relationship between quality of life and the potential for the gut bacterial community to synthesize 3,4-Dihydroxyphenylacetic acid, which is one of the molecules that dopamine is broken down to in the nervous system. Dopamine is a neuro-transmitter, a chemical involved in the functioning of the nervous system, and lower than usual levels of dopamine are associated with depression.

The ability of the microbiome to produce molecules related directly to our nervous system, and more pertinently to produce molecules related to those with a well-characterised part to play in our mental health, is intriguing but at the moment such links remain correlative and not causal.[10] We also know that these molecules can influence the growth of bacteria but we don't yet know if, and how, molecules produced by bacteria interact with our nervous system, how our nervous system might interact with our gut bacteria and whether any of these (currently still speculative) interactions actually affect our risk of developing different diseases. Overall, one thing that isn't speculation is that it certainly feels like a very good time to be a gut bacteria researcher.

Changing our 'in-vironment'

Much of the preceding section has been couched in cautious language because although there is a great deal of support for all kinds of connections between gut bacteria and health, much remains correlative and we lack well-described causal mechanisms. Despite this caution, what is very clearly emerging is an increasingly supported consensus that places our gut bacteria right at the centre of a number of immune-mediated inflammatory conditions. We are also developing more insight into the possible connections between our gut bacteria and depression, connections that are likely mediated through molecules implicated in our general well-being. In many cases, the diseases that are implicated are increasing (for example in the case of depression) and in some cases (for example IBD) those increases in disease prevalence have been firmly linked to 'industrialisation' and 'Westernisation', which are really just other terms for a 'modern lifestyle'. What is it then about our current lifestyle that is having such a profound influence on, first, our bacterial communities and second, through these communities, our immune system?

The first component, changes in bacterial communities, is the more straightforward to understand. If we think of our gut as an ecosystem, then determining the factors that may affect the species living within that system requires an ecological approach. If we want to understand

how an ecosystem in the natural world functions we
have to understand the physical environment and the
ways that species populate and interact with the
environment and with each other. It is only by
understanding these complex interactions that we can
start to understand how communities form, how they
function and when, how and why they might break
down. In a natural ecosystem like a forest, the physical
parameters include temperature, seasonality, rainfall, soil
and pH as well as topography (the 'lumpiness' of the
environment), elevation and aspect. The species present
depend on all manner of factors including how long the
ecosystem has been around in its present form, what
ecosystems surround it, what species managed to colonise
it initially (founder effects) and how colonisation
changed the ecosystem (ecological succession). Our gut
is certainly a small ecosystem, and compared with a
tropical forest it may have a reduced biodiversity (there
are no birds, mammals, reptiles, plants and so on), but it is
still a product of history and interactions and is still
sensitive to perturbations, extinctions and colonisations.
If we think of our gut as an ecosystem, suddenly some of
the most important factors affecting it become obvious:
colonisation, bacterial interactions and nutritional input.

The nutritional environment of our gut bacteria is
determined solely by the food we eat. Nothing else is
there for them to consume, so it makes good sense to
think that our diet might have a very strong part to play
in governing the structure of our gut bacterial community.
We already know from previous chapters that the
'modern' diet incorporates a balance of foods that cause
problems to our health in part through mismatches with

our evolutionary history. Given that we have evolved in tandem with our bacterial communities, and that they have been selected to thrive in a nutritional environment rather different from the one we have imposed on them in very recent times, then diet is certainly a good place to look for mismatches.

Embracing the geographical variation that exists in terms of disease and genetics often gives us the insight we need to figure out exactly what is going on (as we saw in Chapters 2 and 3), and the gut bacteria–diet connection is no different. What we see globally in different populations is a great diversity in gut bacterial communities both within and between populations, and this diversity reinforces the fact that our gut bacteria have evolved *with* us. The Japanese population, for example, has a gene expressed by a gut bacterium that codes for an enzyme called porphyranase. This gene is found in *Bacteroides plebeius* in Japanese people, but it is not found in this same bacterium in other human populations. The enzyme assists in the digestion of seaweed, a component of the Japanese diet that is seldom a major dietary component in other populations. Single gene changes clearly demonstrate the intimate co-evolved nature of our relationship with gut bacteria, but it is patterns in overall gut bacteria communities and potential correlations with diet that are of most interest. Fortunately there are some studies on this topic and they point in the same general direction. A rise in Bacteroidetes species relative to Firmicutes species (so, a shift in the overall community structure of the gut bacteria) has been found in comparative studies of people living in the Venezuelan Amazon, rural Malawi and the USA. *Prevotella*, a Bacteroidetes genus, was favoured in those with a more Western diet that is rich in carbohydrates, but is also the genus that is enriched in patients with rheumatoid arthritis. In general, enrichment of Bacteroidetes and diminishment of Firmicutes are correlated with IBD. Furthermore, three genera of Firmicutes, *Faecalibacterium*, *Coprococcus* and *Dialister*, were reduced in people with depression, while a higher presence of a type of *Bacteroides* was associated with a lower quality of life and depression.

A Western diet seems to be associated with bacterial community shifts towards species that are themselves correlated with inflammatory disease associated with a Western lifestyle. The obvious question to ask then is whether we can shift our communities back to the balance found in those living with diets more aligned with our recent but 'pre-modern' evolutionary history? The answer is a guarded but optimistic 'maybe'. Studies of IBD, for example, do show some benefits of dietary interventions (low-sulphur diets, complex-carbohydrate diets), but at this stage we simply don't know whether any beneficial effects are caused by changes to our gut bacteria community or by changes to some other component of the system.[11] Taking a wider perspective, one study showed that the ratio of Bacteroidetes to Firmicutes increased in people that consumed either a fat-restricted or a carbohydrate-restricted (and low energy) diet for a year. This study was primarily concerned with obesity but the diet of test subjects that showed beneficial changes in bacterial communities were arguably more in tune with our evolutionary history than a modern fat- and carbohydrate-rich diet. We also know that dietary fibre is the critical component of our diet when it comes to nurturing our gut bacteria, and Firmicutes are one of the main responders to fibre in our diet.[12] These species can access the carbohydrates in fibre, but diets low in fibre favour other species that can work on what we give them.

What has been described as the possibly 'dysbiotic' Western microbiome[13] seems to predispose individuals to a range of diseases and although we have much to learn, diet is clearly a contributory factor. The consensus that is emerging is that a modern diet is a mismatch for the bacterial communities we have evolved with and instead favours slightly different communities that interact with our immune system in ways that cause disease. An awful lot more flesh needs to be put on these bones, and we still need mechanisms and stronger evidence for causation, but nonetheless the skeleton is there.

Diet is not the only factor affecting our gut bacteria. Just as animals and plants must find their way to new pinnacles of rock formed by volcanoes in the ocean (like Hawaii, or the Galapagos Islands), so bacterial species must find their way into our gut. This process of ecological colonisation is also affected by novel components of a modern lifestyle that were not part of our evolutionary history, although the strength and importance of these founder effects on our gut communities is hotly disputed.

The most prevalent founder-effect hypothesis concerns something we have already touched on in Chapter 3, delivery method. A number of studies suggest that infants delivered by caesarean section are at greater risk of certain diseases than infants delivered by the evolved method, via the vagina. Of particular interest is the fact that epidemiological studies that examine large-scale patterns of disease reveal links between caesarean-section delivery and increased rates of autoimmune disorders as well as allergies and asthma (more of which shortly). Other studies have linked mode of delivery to differences in gut bacteria. Furthermore, the rate of caesarean-section delivery is increasing globally and that increase is focused in more developed countries, where more than a quarter of infants now enter the world by this method. If these lines of evidence are linked, then the idea develops that vaginal birth provides a 'bacterial baptism', seeding the infant with bacteria from the mother during the birthing process. In contrast, this seeding does not occur in infants delivered by caesarean section and this results in dysfunctional gut bacteria communities and subsequent immune-related diseases. The logic is persuasive enough to have inspired a 'treatment' involving the deliberate transfer of vaginal fluid to the newly delivered infant in a process that has become known as vaginal seeding.

While the logic behind vaginal seeding may be persuasive, the evidence is far less so. For a start, the differences in gut bacteria between vaginal and caesarean-section-delivered infants are transient and disappear after weaning. However, that early period of life, when caesarean section-delivered infants have different gut bacteria, might have long-term effects on the immune system that cause the differences seen in later life between individuals delivered through different means. The problem is that there are a wealth of confounding factors that are horribly interrelated and also contribute to bacterial differences and to subsequent health including antibiotic administration, labour onset, maternal weight, maternal diet and, as we shall see shortly, breastfeeding.

As a recent, and critical, review of the topic concluded, 'numerous studies have demonstrated an association between CS delivery and altered microbiome establishment, [but] no studies have confirmed causality'. The recent discovery that infant colonisation actually begins *in utero* is fascinating but is yet another complication in an already complex story.

The second colonisation process implicated in postnatal colonisation of the gut is again something we have met previously: infant feeding method. Breastfeeding is the evolved solution, but recent technological developments interacting with social factors prevalent in Western societies have led to an increase in the number of infants fed predominantly or exclusively with formula. Numerous studies of infant stools have revealed differences in the bacterial communities of breast- and formula-fed infants, and studies also suggest that breast-feeding confers protection against a number of diseases including, crucially, some inflammatory diseases and IBD.[14] In what is now a familiar refrain, the evidence is mainly correlative and we still lack a solid grasp on the mechanisms involved. However, some evidence is now suggesting that it is breast milk and oral contact with the skin around the nipples that is responsible in part for seeding the infant gut with maternal bacteria.[15] During the first month of life it seems that nearly 30 per cent of infant gut bacteria come from breast milk and more than 10 per cent from the skin surrounding the nipple. But as we've seen with caesarean-section delivery, bacterial community differences are short-lived and longer-term health implications are complicated by an array of interrelated factors.

More mismatching: more problems

If we are seeking to blame the modern world and its mismatch with our evolutionary heritage for our woes, then looking at diseases or conditions that are increasing in developed countries is a good strategy. We have already seen that the observed rise in some inflammatory diseases can be linked via our immune system and gut bacteria to recent changes in our diet and to other 'modern' world developments. Away from our gut, and its obvious connections to our gut bacterial community, there are other immune-related conditions that have increased with the advent and rise of modern living.

Asthma is one such condition. It is caused by an inflammation of the tubes that carry air into the lungs and an attack can be triggered by a range of factors, including allergens such as house dust mites, pollen, smoke, pollution and cold air. That asthma can be triggered by allergens, substances that cause allergic reactions, provides the link to another related suite of conditions that are also increasing in the modern world and are firmly connected to our immune system and the process of inflammation. Allergic diseases and conditions include life-threatening anaphylaxis (an acute and serious allergic reaction to an allergen to which the immune system has developed a hypersensitivity), food allergies (nut allergies being especially common), rhinitis (inflammation of the mucus membranes in the nose, associated with hay fever for example), conjunctivitis ('pink eye', an inflammation of the membrane covering the eye), eczema (an inflammatory condition affecting the skin), eosinophilic esophagitis (inflammation of the lining of the oesophagus), drug allergies and insect allergies (especially wasp and bee stings). Globally, around 300 million people suffer from asthma (this figure is expected to reach 400 million by 2025), around 250 million from food allergies, 400 million from rhinitis and an estimated 10 per cent of the population from drug allergies. These conditions frequently co-occur in the same individual and their burden is considerable. Economically for example, asthma has been calculated to cost close to $20 billion (£16 billion) a year in the USA (2007), and while the problems of having a severe food allergy might be harder to calculate they are clearly constraining and can become the dominant factor in making lifestyle choices.[16] When you can't fly on a plane serving nuts, or eat virtually anything with 100 per cent confidence that it might not kill you, you are hardly enjoying the opportunities presented by the modern world.

For all of the opportunities it presents, our modern world is fairly and squarely to blame for the rise in allergic disease. Increases are strongly associated with urbanisation and affluence, but as always it is slightly more complex than it first appears. For example, large rises have been seen in low- and middle-income countries, so it is incorrect to characterise allergies as somehow being an affliction of the 'pampered rich' (however relatively we want to interpret 'rich'). In many low- and middle-income countries solid fuels, including wood and cow dung, are burnt for cooking and heating in simple stoves or open fires, often with poor ventilation. Smoking may have declined in wealthier countries but it is still prevalent in many low- and middle-income countries, and second-hand smoke in particular is a problem for infants and children, who are tending to bear the brunt of the increases in allergic diseases.[17] These 'pollution' factors are as much as five times more severe in poorer countries and are a contributory factor towards the development of asthma.

The 'hygiene' problem

One obvious evolutionary mismatch that might contribute to the rise in allergic diseases is that we did not evolve to live in confined quarters breathing in smoke. Such an observation is trivial and not especially interesting, but also true. A far more subtle and interesting mismatch has been proposed that seeks to explain the rise by focusing directly on the possible interaction between our immune system and our modern lifestyle.[16] It is an explanation that is so intuitively seductive that it has gained considerable traction and is indeed accepted as fact by a great many people. The 'hygiene hypothesis' is the idea that we are living in such scrupulously clean environments these days that our immune systems, evolving as they did in the dirty 'real world', are never properly challenged and so are unable to learn to distinguish friend from foe. The adaptive 'teachable' component of our immune system develops as a consequence of exposure to potential bacterial friends and foes, and an allergic response is a result of poor learning. In short, in the modern world our immune system has trouble distinguishing between friend and foe and overreacts. Developing this idea into a workable explanation of the rise in immune-related diseases is a little more complex than simply asserting that 'we clean too much'.

Declining bacterial infections in childhood and increases in allergic diseases had first been linked in the 1970s. At the time, the notion that growing up in a farming environment, with an assumed greater opportunity for microbial exposure, could protect against hay fever and allergies was still developing. However, the 'hygiene hypothesis' as we know it really took off much later, after a study by David Strachan in 1989. Strachan was interested primarily in the increase in hay fever and in a *British Medical Journal* paper entitled 'Hay fever, hygiene and household size' he put forward an elegant suggestion to account for this increase, and at the same time, account for the rises observed in asthma and childhood eczema. He wrote: 'Over the past century declining family size, improvements in household amenities, and higher standards of personal cleanliness have reduced the opportunity for cross infection in young families. This may have resulted in more widespread clinical expression of atopic disease [those causing hypersensitive allergic reactions – commonly, hay fever, eczema and asthma] emerging earlier in wealthier people, as seems to have occurred for hay fever.' In other words, unhygienic contact with siblings (and as a father of four I know just how unhygienic these contacts can be) is a 'good' thing in terms of avoiding allergies, despite the resulting infections one might acquire.

This 'hygiene hypothesis' gained considerable traction in the popular press, possibly because it seems so refreshingly straightforward and logical, and possibly because it also allows people to indulge in everyone's favourite pastime; decrying the next generation. Thus, in the 'good old days' children played outside more and came into contact far more with animals, plants and soil. We also lacked the arsenal of antibacterial products that we so assiduously deploy these days. Consequently, we also had plenty of exposure to microorganisms in our 'filthy' homes and our immune systems learnt until they could learn no more. Sure, we may have died from TB and dysentery, but at least we could eat nuts. These days of course everyone is weak, sniffling and wheezing through the streets with their allergies and asthma. The modern immune system is cut off from its bacterial teachers because we are all so hygienic and kids never get to eat dirt like they did back in the day.

Knowing what we do about immunity, gut bacteria and inflammatory diseases it is entirely sensible to suggest that a lack of exposure to bacteria early in life, particularly low-level 'background' or 'sub-clinical' exposure, could cause problems with our immune system's development. That the hygiene hypothesis has biological plausibility is one of its great strengths. However, it's a big step from a seductive idea and some interesting correlations to a scientific consensus on a proven causal relationship.

A central plank of the 'hygiene' concept is that we are keeping our homes *so* clean, *so* devoid of bacterial life, that the resulting domestic dead zone is affecting our children's health. Strachan's paper focused on the inverse relationship between family size and allergic disorders, primarily hay fever. An inverse relationship describes situations where one value increases (in this case hay fever) while the other (family size) decreases. Strachan linked this inverse relationship to the higher levels of cross-infection that are possible in larger families, noting that a reduction in family size (typically seen in developed nations) could result in reduced cross-infection and increased allergies. He also noted that 'improvements in household amenities, and higher standards of personal cleanliness' could reduce the opportunity for cross-infection. These ten words begat the 'hygiene hypothesis' as we now know it, but Strachan did not specifically state that our homes are now so clean that our children's immune systems are not exposed to any bacteria and that's why they have increased allergies. He suggested instead that smaller families were a key factor and then speculated that better household amenities and improved personal hygiene *might* be important. Nowhere in this original paper was 'home hygiene' explicitly mentioned, although clearly one could make that inference. And many did.

A considerable amount of research has examined the different issues that come under the hygiene hypothesis umbrella and the most consistent findings are: that there is a decreasing risk of atopic diseases (particularly hay fever) for people from families with three or more siblings; and that there is a decreasing risk for younger siblings, particularly if older siblings are brothers.[18] The relationship between family size and allergic diseases that was suggested by the original data has been supported by some studies, but the findings are not entirely consistent when individual diseases are examined.[19]

Relationships with family size do not lend any support at all to the notion that our homes are too clean, and neither could they. To do that requires us to examine and measure home and personal hygiene and to quantify the health of the people in those homes. Scientifically, the ideal solution would be to manipulate hygiene in homes, but such a study is unlikely to pass muster at the ethics committee. If we ask people to be less hygienic, we know for a fact that they are likely to suffer health costs from increased potential infections. If we ask them to be more hygienic, then we have some reason to suppose they could end up with more allergies (since this is the basis of the hypothesis we are testing). Another approach, albeit correlative, would be to look at the use of cleaning products and practices and the prevalence of allergies across space (comparing different countries perhaps) and time. As ever, untangling complex social and medical factors is challenging to say the least.

Overall, studies examining the possible link between home hygiene and allergic diseases have pretty much drawn a blank. The link simply doesn't exist. Yes, we are using more cleaning products now than before, but consumption overall or for specific types of products in different European countries has no correlation with the rise in allergic diseases when other factors are controlled for. In a major review of the hygiene hypothesis published in 2006, the conclusion was extremely clear: 'Evidence of a link between atopy [diseases like hay fever, eczema and asthma] and domestic cleaning and hygiene is weak at best.'[5] In fact, they go a little further in their summary, stating that the 'increase in allergic disorders does not correlate with the decrease in infection with pathogenic organisms [those causing disease], nor can it be explained by changes in domestic hygiene'. A second major review of the hygiene hypothesis was published in 2012.[20] I thoroughly recommend reading this technical but approachable review, which is freely available online, if you want to understand more about the hygiene hypothesis and its implications. The authors of this review go one step further, stating that 'The idea that "poor hygiene in itself would be protective" [in other words, that a dirty house protects your children from allergies] is now generally refuted,' adding, one senses wearily, 'although it is still discussed in the popular media'.

The term 'hygiene hypothesis', so firmly centred on home cleanliness, is misleading and unhelpful. These days, scientific papers that mention it usually do so either to suggest that it is no longer used or as a part of some introductory historical scene-setting. However, the death of the simplistic 'clean house' model does not spell the end for the entangled concepts sheltering beneath the hygiene hypothesis umbrella. In fact, the basic concept of a link between microbial exposure and allergic disease is accepted and has become the consensus.

Returning to Strachan's original formulation, the link between allergy and infection is based on the assumption that family size is a realistic proxy measure of cross-infection. Proxy measures are measures of something we can quantify that we take as being representative of something we can't. To make a wider inference, family size could also be a proxy measure for exposure to microbes in general. Scientifically, the hygiene hypothesis rapidly evolved from these early beginnings into the far broader notion that our modern lifestyles cause a decline in our exposure to microbes and a related increase in allergic disease through an impoverishment of learning opportunities for our immune systems. This enhanced 'Hygiene Hypothesis 2.0' embraced exposure to non-pathogenic bacteria including species found in the wider environment, components of bacteria such as the toxins they can produce, and lifestyle issues that reduce microbial exposure, including the increase in urban living, reduction of contact with the 'environment' and with animals, and the decline of familial bed-sharing. In other words, all the trappings of the modern lifestyles we have created for ourselves lead to significant health issues because of a mismatch with our bacterially co-evolved past.

The Karelian peoples of Northern Europe provide an example of the sort of correlative evidence that helps to build the case for a connection between lifestyle (environment in a broader sense) and immunity-related diseases, in this case, the autoimmune disease type 1 diabetes. Ethnic Karelian people living in Russia have very low levels of type 1 diabetes and yet just over the border in Finland, at the same latitude, there is a *six-fold* increase in its incidence. This increase occurs against a near-identical genetic background; in other words the Karelians in Russia and those in Finland are not 'different' ethnic groups. They are genetically the same population, albeit one that has split recently, as a consequence of politics, into two sub-populations that rarely mix. There simply hasn't been enough time to evolve significant genetic differences between these recently divergent sub-populations and yet there is a massive increase in the incidence of type 1 diabetes in the Finnish group. If it's not caused by genetics, such a difference can only be a consequence of a different environment. In Russia, the Karelians live in underdeveloped settlements in relative poverty, whereas in Finland (mostly in a region called North Karelia) the population is modernised and urbanised. A major difference between the two groups is that in Finland the exposure to microbes is far less than in the 'closer-to-nature' Russian population.

Breaking up with 'old friends'

Reconciling what we know of the immune system and microbial exposure with lifestyle changes is clearly a challenge but the 'old friends' hypothesis, proposed by Graham Rook and colleagues in 2004, was an attempt to draw the threads together. Rook and colleagues surveyed the existing work and concluded that the increasing failure of regulatory T cells (at the centre of the inflammatory responses seen in the inflammatory diseases increasing in the modern world) is a consequence of reduced exposure to microorganisms (note, not just bacteria) that have had a *continuous presence* in the environment of mammals throughout their evolutionary history. They call these microbes our 'old friends'.

Immune systems, in us and in other mammals, did not evolve over the last few hundred years. The human immune system evolved within our most distant human ancestors, an evolutionary process that itself built upon the immune system that had evolved in more distant, non-human ancestors. Our ancestral evolutionary environment was the real, dirty, biodiverse world 'out there' and our relationship with it would have been intimate. Contact with mud and soil was inevitable, as would have been contact with the microorganisms dwelling within them, and within the faeces of animals and our fellow humans. These include parasitic worms (flat worms and nematodes) and viruses as well as bacteria. As we gathered leaves and berries and ate them unwashed, and hunted animals that we processed with our bare hands, our contact with microorganisms was frequent, biodiverse and absolutely guaranteed. Against this background of constant potential invaders our ancestral immune system had responded to threats, but it would be inefficient and undesirable to react to absolutely everything. First, such an 'all-in' response comes with a considerable cost to the bearer of that immune system, as we see in those suffering from allergies and inflammatory diseases. Second, as we have seen, components of our symbiotic gut microbiota are absolutely vital to us and were just as important to our evolutionary forebears. Immune systems need to learn to be tolerant to organisms that are so common they can't be avoided, and to those that are beneficial to us. We have evolved with these organisms, these 'old friends', and now we have evolved dependence on them to regulate aspects of our immune system. We need them to teach our immune system to walk the delicate tightrope between too much and too little – without the help of our old friends the immune systems overbalances and falls. It isn't exposure in general, but specific exposure to these old friend microorganisms that is the key.[10]

Island living

John Donne was wrong. We are an island, each of us providing a niche-rich bacterial habitat that, like a volcanic island rising from the ocean, is ripe and ready for colonisation. Bacteria moved on in, set up home and evolved with us in a mutualistic relationship whereby both island and inhabitants benefit. Evolution has honed our immune system and adapted to, and shaped, chemical communication channels between our bodies, our bacterial multitude and other microorganisms to ensure that we can tolerate our friends and respond with appropriate force to foes. The modern world we have created provides numerous and diverse ways in which these delicately poised relationships can break down, from our mode of birth to our diet via family structures and domestic set-ups. In just a few decades, the accelerated development of our lifestyles has resulted in an accelerated increase in all manner of conditions that are either firmly connected to, or looking very likely to be connected to, our gut bacteria. Once again the resounding message is one of environmental/evolutionary mismatch and like the mismatches we've already met, the consequences are serious at individual and population level. The solution? Well, we are seeing the development of faecal transplants, with new bacterial communities directly introduced from healthy to dysfunctional guts but, as boring as it sounds, eating more fruit and vegetables seems like a less drastic first response. Maybe that is the real value of the paleo diet (Chapter 2)?

CHAPTER FIVE

Stress: From Saviour to Killer

Stress is everywhere. We have stressful jobs, get involved in stressful situations and people stress us out. You might have stressful days or stressful weeks. Indeed, many people report feeling stressed constantly. Unless you shut yourself off from all media (and more of that in Chapters 6 and 9) you cannot escape the fact that 'stress' in various guises is making us ill and it is becoming increasingly obvious that our lifestyles are to blame. But how can our comfortable and safe modern lives possibly be more stressful than those our ancestors faced? I haven't wrestled any bears or survived any famines lately and in all likelihood neither have you. Why then does the comfortable, technologically enhanced modern world that we have so assiduously created to make life more bearable and luxurious cause us 'stress'?

The word *stress* is used in all sorts of scenarios when we feel under pressure, rushed or otherwise frazzled by the demands of modern life. However, the true meaning of the term, and its medical implications, is far more subtle than the headlines would have us believe. By examining what stress actually is, as opposed to what click-bait headlines might have us believe it is, we can start to understand stress, how it's affecting our health, how and why it evolved and how our modern lives interact with this evolutionary heritage in such negative ways.

Stress is a killer...

The implications of stress in the modern world go far beyond intellectual curiosity over evolutionary mismatches. Stress is a killer. It is possible for acute stress to cause heart attacks, as shown by a rather elegant if morbid 'natural experiment' in Athens following a large earthquake in 1981. Researchers examined records of deaths in the days following the quake on 24 February 1981 with records of deaths that occurred in a similar period in the year before (1980) and the year after (1982). They found more cardiac-related deaths than expected in the days after the earthquake but no effect on deaths related to cancer, and little if any increase in deaths from other causes. The 'excess mortality' identified in 1981 was more evident in people with underlying heart disease and the clear conclusion was that stress caused by the earthquake had triggered heart attacks, especially in those with conditions that left them vulnerable.[1] Sudden death from cardiovascular problems has been shown in the aftermath of other quakes in Australia and California, underpinning one simple fact: acute stress can be fatal.

Acute stress and heart attacks are at one end of the stress spectrum, and the fact that you could die at times of acute stress perhaps comes as no surprise. The increase in the cardiovascular diseases that often underlie such stress-triggered mortality are caused by our modern diets combined with a lack of exercise playing against a background of genetic susceptibility to such diseases in many people. This is a theme becoming familiar from earlier chapters and it shows an obvious mismatch between our evolved physiological responses to sudden, potentially dangerous events and the problems that a modern lifestyle causes for our cardiovascular health. It is, however, the medical effects of long-term, chronic stress that are especially worrying from a public health perspective and that are most affected by our recent lifestyle shift interacting with our evolutionary heritage. A constant drip-feed of relatively low-level physiological stress in response to pervasive 'micro-stressors' (a groaning email inbox, a missed bus, financial worries, children, relationship problems, an empty fridge, interest rates, climate change, Instagram, the laundry…) causes, as we shall see, the release of a range of chemicals within the body that collectively and cumulatively cause problems over the long term. Common physical effects of stress include headaches, muscle tension and pain, chest pain, fatigue, loss of libido, upset stomach and problems sleeping. Each of these physical effects cascades through to our mood and our behaviour and that in turn can create more stressors in an immensely harmful feedback cycle. Stress-induced headaches make us anxious, anxiety can trigger eating disorders, and pain causes restlessness, mood swings and self-medication through drug and alcohol abuse. Meanwhile, fatigue and sleep problems cause further anxiety and a sense of being overwhelmed that can lead to social withdrawal and yet more self-medication via the off-licence (Chapter 8). These spiralling stress symptoms have less immediately noticeable effects but they have been implicated in causing diseases like depression, Alzheimer's and even, at least according to some headlines, cancer (but read on to see why those headlines can probably be ignored). As we shall see though, disentangling the potential connections between modern lifestyles, micro-stressors, disease and human evolution is far from simple. Before we get to that we must first answer a very important question – what is stress?

... But it keeps us alive

Animals, including us, like to operate in a state of constancy. We don't generally like change and we especially don't like major change that happens rapidly. The process by which a steady, favourable state is maintained in our bodies is called homeostasis and most of the time we can easily keep everything ticking over exactly as it should. Think of it like a thermostat turning the boiler on and off to regulate the temperature in your house. We don't have to think about it and most of the time everything works smoothly.

Physiologically, stress is our response to an environmental challenge that pushes us outside of our normal, stress-free regulation; perhaps we fall into icy cold water or a tiger suddenly appears from behind a bush. When the environment changes suddenly, and for the worse, we must do something about it and do something quickly. Our normal processes simply aren't going to work any more, and sitting back and hoping that things will magically return to normal is not going to fix the immediate and serious problem. We need to make things happen and the way we do this is by 'stressing'. Stress is, at its most basic level, an evolved biochemical response to danger. And it's very useful.

Fight or flight

We have a wonderful response to scary and life-threatening situations that operates entirely outside of our conscious control. It can cause us to wet ourselves, go pale and start shaking but, despite these unpleasant side effects, it can also save our lives. It is called the *acute stress response*, or more commonly 'fight-or-flight'. In essence, our bodies get pumped up and ready to fight or flee. However, the details are more complicated than 'we get pumped up' and they are worth exploring because the more we understand how the fight-or-flight response works, the more we can start to appreciate the medical problems related to our stress response in the modern world.

Stress begins in the brain. Made up of a number of different discrete but interconnected physical structures, the brain is complex. It includes relatively well-known structures like the cerebellum, the hippocampus and the cortex as well as less well-known structures like the left and right amygdala (from the Greek word for almond, which they resemble) located deep in the brain. The amygdala are concentrations of nerve cells that are very closely linked to our emotions including happiness, sadness, anxiety and fear. It is here that sensory information, especially from our ears, eyes and sense of touch, is processed very rapidly and without our conscious input. If a threat is perceived, the amygdala begin an impressive and complicated set of reactions that cascade throughout the body. The reaction begins with a 'we're under attack!' nervous signal sent to another part of the brain, the hypothalamus.

The hypothalamus doesn't do much 'thinking' but instead acts as a link between our brain and the chemical messengers, hormones, which circulate in our bloodstream and control much of what goes on in our bodies.

When we get scared, the hypothalamus kicks into overdrive and activates the sympathetic nervous system. Operating entirely unconsciously, the sympathetic nervous system's primary job is to coordinate our fight-or-flight reaction. Once stimulated by the hypothalamus, it causes us to speed up, tense up and become more alert. These are great responses to an immediate danger. They prepare us to kick some ass or, as is usually the best plan, to run away.

As well as activating 'superhero mode', the hypothalamus sends out signals to turn off the muscles moving food through our digestive system (no need to waste energy on systems that won't save our life) and to make blood vessels contract (hence we go pale). At the same time, nervous signals stimulate our adrenal glands, pyramid-shaped organs that lie on top of our kidneys. Once the adrenal glands are stimulated they start rapidly producing adrenaline and noradrenaline (and we are talking milliseconds here since we have no time to lose). This pair of 'stress hormones' causes, among other effects, our heart rate and blood pressure to increase rapidly. We tend to think of hormones as 'slow', perhaps because we commonly think of them in relation to things like puberty, sex drive and the menstrual cycle, and nervous signals as 'fast' and although that is technically true (nervous signals are conducted through the body far more rapidly than chemicals can be circulated) the endocrine system is far from sluggish when short distances are concerned.

Meanwhile, the hypothalamus is messaging the pituitary gland to release another hormone, adrenocorticotropic hormone (ACTH), that again passes rapidly through our bloodstream and stimulates the adrenal glands even further. This additional stimulation causes them to release a heady mix of yet more hormones, including one that will crop up quite frequently in this chapter, cortisol. All of this frantic activity serves but one purpose: to hype the body up into a state where we can get ourselves out of an immediately life-threatening situation. Make no mistake, seeing that tiger jump out from behind a bush causes profound changes to our body, and thanks to evolution we might just make it out alive. It is probably fair to say that this remarkable and complex interaction between the nervous and endocrine (hormone-producing) systems kept our ancestors alive long enough to have the offspring that form the chain all the way to us. That 'stress-thankful' chain could be very long indeed. Some kind of fight-or-flight response mediated by adrenaline and related hormones is found in a remarkably wide variety of animals, including reptiles and amphibians as well as invertebrate groups like molluscs and insects,[2] neither of which form part of our lineage but you get the idea – stress responses are more or less ubiquitous.

The harmful effects of stress

Words that have a precise meaning scientifically or medically, but are also used in broadly similar ways in everyday life, are hugely problematic for scientists and non-scientists alike. A fine example of this is the seemingly innocuous word 'theory'. In everyday language, 'theory' promotes a sense of unsureness, of speculation, or of some ivory-towered notion that is unlikely to translate to the real world. 'It's all theoretical' is almost an insult, suggesting some naive and unworkable idea. In science on the other

162 UNFIT FOR PURPOSE

hand, theory means quite the opposite. A theory is a robust and evidenced framework to explain something that is based on a set of well-supported and interlinked principles. The use of this term in connection with evolution in particular causes huge problems. Those who seek to undermine the foundations of modern biology will often stand on the side-lines hurling in the phrase 'but it's only a *theory*', as though that somehow negates the huge deluge of evidence that is contrary to their deeply held belief. Likewise, stress has multiple meanings that go beyond the biomedical neuro-hormonal response to danger. Initially, stress was used as a medical term because of clear analogies with the stress (and also 'strain', another word with medical and everyday meanings) undergone by materials like a bridge girder as forces are applied. Even before we had the pathways involved in the stress response nailed down though, and before stress had a precise defined biomedical definition, doctors were clearly aware that the pressures of life had some implications for health. In a wonderfully titled paper from 1919, 'The influence of psychic acts on the progress of pulmonary tuberculosis', a Japanese medical scientist called Tohru Ishigami concluded that anxiety ('psychic acts', or as we would now call it, stress) caused by an overtaxing education system was causing a high mortality among young people with tuberculosis. In other words, children with TB were dying not because of the TB but because of the additional burden that TB put on their systems as they attempted to navigate the stressful Japanese education system.[3] This startling conclusion, a warning perhaps for those with interests in education policy, was no flight of fancy. Ishigami, in an integrative, cross-disciplinary study that was well before his time, linked together laboratory research on guinea pigs, rabbits and rats, and epidemiological data on tuberculosis in Japan with developing research by Walter Cannon. Cannon was the scientist who invented the term homeostasis and also coined the phrase 'fight-or-flight'. By combining some of Cannon's early work on the biochemistry of the stress response with his own findings, Ishigami was able to start to piece together and produce evidence for the influence of stress on our health. Ishigami started on a quest that has continued apace over the century that has followed.

The harmful aspects of the stress response can be divided into acute and chronic. We've already seen the effects of acute stress in the Athens earthquake study. The sudden deluge of stress hormones triggered by emotional or physical stress causes increased heart rate and blood pressure, and in those with a healthy heart these increases enhance our ability to fight or flee. A good thing in many people, of course, but in those people with underlying conditions affecting the heart, such as coronary heart disease (where the blood flow supplying the heart muscle is restricted), the sudden increase in demand placed on the heart by stressful situations can prove fatal. For this reason, in addition to the usual 'sensible quartet' of not smoking, eating a healthy diet, drinking alcohol in moderation and exercising, medical advice now acknowledges the problem of stress and makes suggestions that even 10 years ago would have felt slightly more Glastonbury than Harley Street. As well as the rather obvious advice of 'Avoid situations that you know will cause stress' comes, for example, recommendations that 'Quiet time, meditation, prayer, reading, yoga, and relaxation techniques can help in stress management'.

The 'typical' modern life

To appreciate the role that chronic stress plays in many of our lives we need to consider a 'typical' modern life. In particular, we need to examine closely those aspects of our lives that can be categorised as 'modern' and consider whether these aspects are resulting in a mismatch between our twenty-first-century lifestyle and our past evolutionary environment.

Defining a typical modern life isn't going to be easy because we all live different lives. If we take a broad-brush approach though, it is possible to assemble some of the components common to many people's lives and thereby come up with a typical day that includes at least some of the stressors present now that would not have been an issue to even our reasonably recent ancestors. Given we cannot go back in time, we are faced with a slight problem because we need a suitable group with which to compare our typical modern life. It is tempting (indeed, I have already been tempted in Chapters 2 and 3) to take extant hunter-gatherer tribes as our comparison group. Although this is a sound approach in some respects, it would be a mistake to assume that somehow these groups represent modern humans 'minus 12,000 years or so'. The Sān people of Namibia, for example, are very much modern humans and although their lifestyle is more representative of a 'simpler', pre-modern lifestyle that does not mean that the people themselves represent some pre-modern human. It is possible to tie yourself up in knots over this type of comparison, and many have, but as we'll see my analysis of a modern lifestyle gives a rather easy escape from an argument that can start down a path no one wants to tread. That get-out-of-jail-free card is provided by the simple fact that many of our modern micro-stressors are either a direct or an indirect consequence of technology that wasn't (and isn't) a part of a pre-modern lifestyle, whether that is one lived now or 12,000 years ago.

For the purposes of determining some of the stresses of modern life, I am going to bundle together differences between men and women to create a single 'human' lifestyle. I know this is problematic but it is also worth saying that as well as sex, stressors will differ with age, where you live, your background, your job and many other factors; detailing everyone's lives is going to be, well, too stressful. With these caveats and cautions in place, let's begin to map a typical twenty-first-century 'modern lifestyle' day. If your day is different from this, and I sincerely hope that it is, then with any luck you will still recognise some of the themes that develop.

6.30 a.m. Wake up 30 minutes before the alarm goes off worried that the alarm won't go off because your phone will have decided to update overnight and wipe your alarm settings when it restarts. Or the phone will have mysteriously consumed almost all of the 100 per cent battery charge you gave it yesterday evening. Ponder whether you should keep the phone plugged in all night despite the small fire risk and some complex information about battery 'memory' that you half-read yesterday. And if you have the sound turned down for notifications and calls does that affect the alarm? Drift quickly back into a far deeper sleep than you managed during the night.

7.00 Woken by an annoying and loud alarm from a dream involving being unable to dial a phone number correctly, the consequences of which are too ghastly for your now partly awake brain to formulate in any sensible way. 'How much sleep did I have? Was it enough? Am I getting the seven to eight hours recommended by the latest Facebook post that I'll be reading in approximately one minute...'

7.01 Turn on phone. Spend the next 15 minutes blearily 'catching up' on 78 WhatsApp messages from a group of friends arguing over which Star Wars film is best, social media notifications involving the birthdays of some people you don't actually know and 'BBC Most Read' news items. Learn that there has been a hurricane that has killed hundreds of people in a place you have never heard of and that a family you've never met has died in a fire in a city that you've never visited. Find out that a celebrity you have never heard of who is famous for being in a 'scripted reality show' (seriously, a what?) you have never watched has split up with someone else you don't know who is largely famous for having sex with someone in another reality show you didn't watch. Despair at the state of the world and end up mistyping a tweet in which you mean to say that the 'world is shallow and superficial' but somehow end up typing that the 'worm will swallow and is super special'. Copy the tweet to avoid retyping the rest of it, then delete the tweet and then repost the tweet correctly after three attempts. There is still a typo.

7.20 Children wake up. At this point, if you have children you will understand and if you don't then you won't. And of course society gives you plenty to stress about if you don't have children because apparently that is now everyone else's business. Why don't you have children? Do you want children? Why don't you want children? Are you trying hard enough to have children? Have you eaten enough zinc? Is your childless situation selfless or selfish? There's really no let-up.

7.20–8.30 A terrible blur of breakfast during which you can worry that the cereal you are pouring is too sugary and lacks nutrition, that the milk you bought at a rock-bottom price is destroying farmers' lives, that the bread you are eating has come from an agricultural system that is crippling the environment and contributing to the sixth great extinction, or that the one beacon of hope in your morning, a cup of freshly brewed coffee, is going to give you cancer. You decide to check the latest findings on the whole coffee-cancer thing on your phone and wish you hadn't. You brush children's hair, a terrible experience for both brushed and brusher. About to leave, late, you realise teeth haven't been brushed. Cue a massive rush to brush teeth so that tooth decay doesn't set in and ruin your children's smile for ever.

8.35 Run to car. It's low on fuel (that is causing climate change of course) but, hey, there's got to be enough in there for a few miles, right? There's a weird rattle… turn up the radio to block it out, and of course you can hear more about that hurricane destroying thousands of lives. Great.

8.40 Be informed by all-knowing children in the back that it is World Sausage Day and that the school is having a 'dress like a sausage' fancy dress day, a fact apparently conveyed to all parents the week before via a message on some online forum that you are literally hearing about for the first time right now. Oh, and everyone needs pound coins for the charity donation that is required for fancy dress. Of course they do, because it's not enough to pay taxes and give to charities you support, you also have to donate to charities you don't necessarily agree with so that your children can take part in some activity that doesn't in all honesty seem to enhance their education. You realise that what started as an inner monologue is now very much an outer monologue that is beginning to unnerve the children.

8.45 Somehow you improvise a serviceable fancy-dress outfit from the detritus that has collected in the boot of the car. A victory for untidiness. Likewise, you scrabble together enough change to fool the casual observer that it covers the donations required, when in fact it is 7p short, not including two foreign coins and a washer that may or may not be part of the seat-reclining system.

8.50 Drop off children at school, an operation that requires a mixture of precise but hugely aggressive driving and split-second door-opening accuracy. It is not for the faint-hearted...

8.55 Fuel light now on. Pull in to petrol station. Ignore signs and check email on phone while gesturing to an attendant, who is achieving the impossible feat of simultaneously both looking and not looking at you, to reset the pump from the previous customer. Eventually fill up with fuel. Decline three increasingly desperate offers to upsell your petrol, including a chocolate bar the size of a table top and a multi-tool that will break on its first deployment.

9.00 Email alerts start tumbling in.

9.30 Arrive at work to spend an hour and a half doing little but answering emails in a futile attempt to achieve the mythical 'zero inbox'. Steadily come to realise that email is a beast of its own making and that the only way to stop it is not to send any. Continue to send emails...

11.00 Sit in a meeting that you know from the start will achieve nothing. Minutes are taken; hours are wasted.

13.00 Meeting ends having achieved nothing. You go to the canteen and after examining the caloric value of all the sandwiches you opt for a simple ham and cheese. With a sinking heart you realise you are shelling out more than 20 times the actual value of the ingredients and you make yet another pledge to make your own lunch, a pledge you know will never be honoured.

13.30 Slink off to gardens next door to grab a quick image for your Instagram account which you largely curate to show dreamy images of flowers and trees. You aren't sure why but you get a little zing every time someone 'likes' a post.

13.40 Upload a fabulous image #nofilter. Wander back to work with the notifications already pinging.

14.00 Get call from school. Child has vomited. Marvellous.

14.45 Arrive at school wishing child could have waited to vomit until after school thereby allowing you to circumvent the 48 hours of childcare during the day you are now going to have to find. You hang around in the car park with a child smelling faintly of vomit until the others emerge.

15.17 Second child vomits in car in response to the vomit smell. Third child vomits in response to the sound of the second child vomiting. You now have 'a situation' that will have to be dealt with, ideally using bleach and heavy gloves but the reality is that it is physically impossible to remove vomit chunks from within and beneath the rear seat of a car. It would be easier to recreate Jurassic Park in your garden.

16.00 Receive text message to inform you that the insurance payments you think are going out monthly have not been and you now have 24 hours to sort out your direct debit or you are driving uninsured.

16.30 Eventually get through to the insurance company to sort out direct debit. Hastily put together a shameful dinner of chicken nuggets and chips, with some peas providing a veneer of healthy eating and good parenting. Read some Facebook posts and argue with someone on Twitter while you cook. Answer three work-related emails and make arrangements to work from home for two days while dealing with low-level guilt about the extra work this will mean for someone else and fear about the implications for your weekend workload.

17.00 Your TV subscription has apparently expired. The new bank card you had to order when the old one got lost (later found) has a different number and so none of your automated payments now work. At least that only takes 90 minutes to sort out.

18.30 Commence 'bedtime'. Like breakfast in reverse but with the added frisson of very tired children and hungry cranky adults. Phone constantly buzzing in pocket with notifications from Instagram, Twitter, Facebook and WhatsApp. Realise halfway through the process that you still haven't made two phone calls you've been meaning to make for a week.

20.00 Too tired to cook dinner, you get in a takeaway, the enjoyment of which is ruined by a combination of financial and health-related guilt. Eat food in front of TV while reading internet pages relating to the TV programme you are watching: 'He *was* in *The Good Wife!* I knew it.' Fight a nagging sense that you really aren't following the plot line as closely as the writer probably intended.

23.30 Go to bed, already destined not to get close to eight hours' sleep. Forget to turn off 'notifications' on phone.

2.00 a.m. Woken by a series of notifications relating to a conversation you aren't a part of, which is being had by a group of people who are in the same bar as each other but in a different time zone from you.

Perhaps that typical day is slightly exaggerated, although in all honesty much of what I've outlined is semi-autobiographical and not too wide of the mark for many days I've had. It would be ridiculous though to suggest that our ancestors didn't have some daily stresses that were more trivial than imminent death from a marauding predator or some natural disaster. The evolution of agriculture and the consolidation of populations into more concentrated urban centres would certainly have created and amplified a whole suite of potential stresses that begin to feel like 'modern' problems. Difficult relationships, family politics and social interactions are not recent inventions and would certainly have been exacerbated by centralised urban living. The ability to cope successfully with such potentially stressful social situations would certainly have been an advantage to our ancestors. Indeed, the evidence from primates shows that the ability to navigate social situations is a feature of some primate brains and arose long before the evolution of modern humans.

Likewise, there is no reason to assume that what we might consider to be 'modern' feelings, like guilt, shame and anxiety (more of which shortly), or feeling the weight of others' and your own expectations did not feature in the emotional palette of early humans and that these feelings were not enhanced or at least made more significant in our daily lives as we developed from a hunter-gatherer lifestyle into one more focused on city states, production and 'civilisation'.

The signature of the modern world

What is so characteristic of the modern lifestyle for many of us is the presence of a daily series of almost continuous micro-stresses that have only come into existence very recently, arguably in some cases in the last five to ten years (for example, stress caused by social media or 'phone addiction'). Physiological stress, as defined by the consequences of the hormonal and neuronal mechanisms we met earlier, evolved as a means to deal with situations that require 'fight or flight'. A squirt of adrenaline and some raised blood pressure and heart rate is ideal if you lose your footing while out collecting berries or during the final stages of a hunt (these are after all potentially existential crises), but is clearly overkill in response to a 'difficult' email from your boss. Despite that, our basic biological set-up is largely unable to tell the difference and so, if we are prone to getting stressed, we can end up feeling stressed almost constantly. Technology has changed our environment far quicker than we can respond either socially (through established rules of engagement for example) or evolutionarily. The resulting mismatch between our physiology and the environment is great news for purveyors of relaxation courses and spa breaks, but it is bad news for our health.

The cumulative effects of chronic stress are well documented. Experiences of prolonged or recurrent stressful events in daily life, especially those which are seemingly (or actually) beyond our control, result in the activation of the stress response and the release of stress-related hormones. Producing these stress hormones in an active attempt to maintain homeostasis (and thereby prevent damage) in the body has been termed allostasis. The flip-side of this hormonal balance, the wear and tear that can be caused by prolonged periods of elevated levels of these hormones, has been called allostatic load. Our total allostatic load is the combined cumulative strain that chronic stress, exacerbated by a modern lifestyle, puts on various systems including cardiovascular, metabolism, immune, endocrine (hormonal) and nervous systems. Thus stress itself isn't an illness but it can cause serious illness if it is allowed to continue. This is because the multi-system damage that is caused by the chronic presence of stress hormones makes us susceptible to health issues that include high blood pressure and subsequent heart damage, diabetes, rheumatoid arthritis, suppression of our immune system and mental health issues.

Stress and mental health
Decline in mental health marked by an increase in depression, anxiety and other disorders is undeniably one of the major medical trends in recent years.[5] In the UK, the National Institute for Health and Care Excellence (NICE) lists as common mental health

disorders depression, generalised anxiety disorder, social anxiety disorder, obsessive compulsive disorder and post-traumatic stress disorder (PTSD), and most of these show a sharp increase in the number and proportion of sufferers in recent years. For example, the proportion of people suffering from generalised anxiety disorder rose from 4.4 per cent in 2007 to 5.9 per cent by 2014, depression from 2.3 per cent to 3.3 per cent and OCD from 1.1 per cent to 1.3 per cent. Set against a UK population of around 65 million people, these disorders account for millions of people suffering from conditions that we know can be triggered by, and greatly exacerbated by, chronic stress.

Of course, apparent increases in certain disorders might be because of a greater recognition of those disorders, both by sufferers, who may be likely to seek help for symptoms that are linked to named and well-known disorders, and by medical professionals, who may be more likely to diagnose such disorders. The increase in prevalence of a disease or disorder because of an increased tendency to diagnose it might be expected to be especially great in mental health disorders in recent years. We have undergone something of a social epiphany lately in recognising the importance of mental health, and the public discourse on this topic is far more open and inclusive than it ever has been. That both Princes William and Harry, for example, feel able to speak openly of their emotions and feelings triggered by the death of their mother, and are able to place those experiences within the framework of mental health, shows how far we have come in recognising that the pressures and stresses of the lives we lead can cause serious problems. A similar background of celebrity-led awareness of mental health issues has been seen elsewhere with Ariana Grande, Justin Bieber and Lady Gaga all opening up publicly in the USA about mental health pressures.

The social background of increased awareness and openness about mental health issues is a problem when it comes to interpreting trends. It is possible though to begin to tease out the different factors that might account for a rise by comparing different sectors of the population in carefully planned and conducted studies. An examination of the prevalence of depression in the USA and its possible link to substance abuse, for example, concluded that depression had more than doubled in the 10 years between 1992 and 2002, increasing from 3.3 per cent to 7.1 per cent. Their structured approach allowed them to show that there were significant increases across age ranges and the increases were irrespective of whether people were white, black or Hispanic. Their initial hypothesis, that depression increases could be linked to substance abuse, was only supported for black men aged between 18 and 29, but they did acknowledge that increased awareness of depression via media campaigns and the advertising of antidepressants may have played a role in the overall increase in prevalence. As they point out, 'These campaigns began during the interval between the two surveys, and just such a general factor is consistent with the broad-based increases seen in this study.' That point notwithstanding, their overall conclusion was still that depression had increased in prevalence and that this increase, if it continues, is likely to have profound effects on public health and on the economics of healthcare. Their overall conclusions are echoed by a great many studies on the epidemiology of depression and anxiety disorders; these disorders are increasing, and this is a problem. What is more, there seems to be good general agreement both in the scientific literature and in the webpages of reputable health advisory bodies like the NHS in the UK and the Mayo Clinic in the USA, that lifestyle stresses are a significant contributory factor.

That many authorities reach the same conclusions, both on the rise of mental health issues and the cause for that rise, still doesn't remove the possibility that we are seeing the influence of an increased tendency to seek help for, and be diagnosed with, disorders like depression and anxiety. A tendency to rush into a diagnosis of 'depression' for what might otherwise be termed 'normal feelings of sadness' was explored in the book *The Loss of Sadness: How Psychiatry Transformed Normal Sorrow into Depressive Disorder* by Allan Horwitz and Jerome Wakefield.[6] In it, they discuss the idea that the boundaries of depression, and thereby the diagnostic criteria that are used to diagnose it, have broadened considerably since 1980 when the American Psychiatric Association's hugely influential *Diagnostic and Statistical Manual* changed from classifying mental health disorders based on potential causes to an approach based on symptoms. Horwitz and Wakefield argue that this resulted in a far more inclusive definition of depression that brought 'normal sadness' and short-lived unhappiness, which might be rightfully regarded as part of the normal human experience, into the fold of clinical depression. While this may well account for some of the increases in depression (and other disorders) observed around the world, the link between modern lifestyle, stress and subsequently impaired mental health is widely supported and it has become an important part of our understanding of mental health in recent years.[7]

The evolution of emotion

Understanding that stress can be a causal factor in mental health is one thing, but to piece together the evidence that this is somehow caused by an evolutionary mismatch requires us to put together an evolutionary argument. It is entirely appropriate that Charles Darwin can provide us with the beginnings of just such an argument. Darwin was the first to suggest that human emotions, and in particular the facial expressions associated with emotions, were evolved traits that adapted over time in response to selection. As is so often the case with Darwin, subsequent research has largely confirmed his thinking. We have already seen the importance of the developing agricultural environment in shaping our recent physical and physiological evolution when it comes to nutrition, and there is no reason to assume that it didn't also play a role in shaping our stress responses, both physiologically and psychologically. One emerging theory on the evolution of guilt, shame and anxiety (more and more associated with the stress of a modern lifestyle) is that these emotions evolved as a response to the combination of cooperation and aggression that allowed us to become so dominant globally (more on the role of violence in our past and present lives in Chapter 7). The theory, advanced by Peter Breggin, supposes that the potent combination in our early evolution of close-knit social groups and a tendency for violence would make it extremely difficult to live together without destroying each other.[8] The solution was the evolution of emotional restraints against 'aggressive self-assertion within the family and other close relationships'. Thus guilt, anxiety and shame may have evolved as mechanisms to counter our violent tendencies. While undeniably useful, evolved emotional restraints against violence have now become constraints, with guilt, shame and especially anxiety becoming socially and emotionally crippling in the modern world. In many cases, the trigger for anxiety is stress, often chronic but also acute. Anxiety in particular has created a veritable ecosystem of different recognised disorders including generalised anxiety disorder, panic disorder, obsessive compulsive disorder (OCD) and post-traumatic stress disorder (PTSD).[9]

The link with stress is apparent quite clearly in the name of PTSD, but it is important to emphasise that not all anxiety (and not all depression for that matter) is a consequence of the sort of stress our modern lifestyle places on us. There are a wealth of reasons why a person might develop an anxiety disorder, and we are still at the start of understanding the complexities of depression, its causes (which, as discussed in Chapter 4, may even be associated with our gut bacteria) and its treatment. Nonetheless, we do know that chronic stress causes elevated levels of cortisol, the 'stress hormone', and reduced serotonin. Serotonin is sometimes called the 'happy chemical' and although it is closely linked to mood and anxiety, it also has a series of complex roles in aspects of our biology that are themselves linked to our mood, including appetite, eating behaviour and sexual behaviour.

Thus, we can certainly make some very solid links between the biochemical products of stress and symptoms related to anxiety and depression, but we are still pinning down the precise connections.

Other brain-related problems that are increasingly linked to stress are Alzheimer's disease and dementia. Studies of mice have shown that provoking modern life-like emotional and physical stresses on mice severely impaired memory and may exacerbate the sort of cognitive defects found in the very early stages of Alzheimer's. If mice aren't convincing enough for you (and as useful as they are in biomedical science, they shouldn't be that convincing), a major review of the literature on stress and dementia risk in humans concluded that stress could indeed play a role in the risk of dementia but that it is part of an array of risk factors. It is not surprising perhaps that stress might play a part, since we know that the immune system takes an important role in the development of dementia, and we know that stress hormones affect the immune system. It is also not surprising that disentangling the connection between lifelong chronic stress and later-life dementia is far from straightforward, and it is very hard practically to investigate a lifetime of stress in any sort of robust scientific manner. However, studies are under way and given the increasing focus on dementia generally, and Alzheimer's in particular, it seems likely that we will uncover both the scale of the risk and the underlying mechanisms in the near future.

Another health implication of stress that is under-standably concerning is the potential link with cancer. However good such a connection may be for headlines though, the science is not supportive. Cancer Research UK is candid in its appraisal of the evidence, stating plainly and unambiguously that 'Most scientific studies have found that stress does not increase the risk of cancer.'[10] Others have been plainer still, dismissing fears that stress causes cancer as a 'myth'.[11] While some small-scale studies have suggested links, for example with breast and gastrointestinal cancer, larger long-term studies and meta-analyses of multiple studies have consistently found no link between stress and the four most common cancers in the United Kingdom (bowel, lung, breast and prostate). Behaviours associated with stress, like drinking alcohol and smoking, poor diet and a lack of exercise (which may also help to promote stress) can and do lead to cancer, but in this case the connection with stress is indirect. The current lack of evidence for a direct link is in one sense reassuring but, if we consider what we already know about the mechanism of stress and the multiple systems that stress affects, it is also surprising. Long-term release of stress hormones can induce DNA damage and affect DNA repair as well as weaken and dysregulate the immune system, all of which are implicated in the development and progression of cancer. It does not then seem unreasonable to suggest such a link could exist and while we don't yet have convincing evidence, I am certain that we will see much more research into the connection between stress and cancer (most likely mediated through our immune system) in the coming years. My money would be on Cancer Research UK having to change its website text at some point in the future, but for now we have to say that the evidence is lacking.

The evolution of coping

Further complicating the issue of chronic stress and our understanding of its effects is the fact that the accrual of damage across multiple systems that is caused by chronic stress increases as we age but actually plateaus at around 60. This sets up a situation whereby the damage caused by stress in early life is only manifested in much later life. Of course, from an evolutionary perspective this is not necessarily good news for our youthful well-being in the long-term. This is because negative consequences of behaviour that only occur after we have reproduced are more or less invisible to selection. However, although Alzheimer's and heart disease brought on by chronic stress might only manifest in later life, there are many other symptoms of chronic stress that manifest far earlier and certainly do have a profoundly negative impact on our quality of life and, by inference, our reproduction. If you had a genetic tendency to be able to handle modern-life stress then some of the early manifested side effects, like a reduced sex drive, depression and insomnia, which would arguably reduce your ability to reproduce, would be avoided. Assuming that the all-important genetic basis for 'stress immunity' is present (without which evolution cannot proceed), then it is possible that if stress has a sufficiently large negative influence during our peak reproductive years we might see selection for physiologies that are better able to handle stress. Evidence suggests that this may well have already occurred.

We have seen that acute stress can exacerbate or trigger underlying conditions such as heart disease (as seen in the Athens earthquake study) that we know have a genetic component. However, there may be a more direct genetic link with our ability to handle, or not handle, stress. One key gene implicated in this link is the gene that codes for an enzyme called catechol-O-methyltransferase, known more conveniently as COMT. COMT catalyses the breakdown of catecholamines, a category of neurotransmitters. We have met catecholamines several times already because this family of compounds includes adrenaline and noradrenaline as well as dopamine. It is dopamine in the brain that is the primary target for COMT. The gene coding for COMT has a naturally occurring variant in its genetic code. The variant, when it is translated by our cellular machinery and used to assemble the COMT enzyme, substitutes the amino acid valine that should be in position 158 for methionine. Not such a big deal you might think, and in some cases the odd amino acid substitution in the chain of amino acids that make up a protein makes little difference to the functioning of that protein. In the case of the 158 substitution in COMT, the different properties of valine and methionine and the subsequent differences these amino acids confer on to the final enzyme make a very big difference. The valine version of COMT is more than four times more effective than the methionine version in breaking down dopamine.[12]

COMT is used preferentially in the prefrontal cortex of the brain, accounting for 60 per cent of dopamine breakdown occurring there.[13] Located right at the front of the brain, the prefrontal cortex is involved with complex thought, planning, decision-making, moderating social behaviour and key aspects of personality. It is here that the different COMT variants show their effects. Under conditions where dopamine release may be high (for example stressful situations), individuals with the valine variant perform better and feel less pain. This variant may be particularly useful in threatening environments where peak performance is an advantage despite stressors like the imminent threat of death or pain. As a consequence, this variant has become known as the 'warrior' genotype. In contrast, the methionine version allows bearers to perform better in memory and attention tasks, making these people thinkers rather than fighters. Perhaps because of the pleasing rhyme and alliteration this variant has become known as the 'worrier' genotype.[12] 'Fighter' and 'thinker' feel like less pejorative titles but 'worrier' links nicely to the idea that increased stress leads to anxiety and to further issues. The 'worrier' genotype seems to have evolved more recently. Its evolution is suggested to reflect selection in more complex environments where maximal performance on complex attentional and memory tasks may have been more important for survival and reproduction than a more straightforward 'hit it and run' strategy. The persistence of both variants (an example of a polymorphism) is likely a reflection that both forms have a selective advantage depending on the environmental circumstances; sometimes it is better to be a warrior and sometimes it is better to be a worrier.

The drip-drip of modern stress might seem to favour the warrior genotype in terms of the health benefits of dealing with stress, but the complexities of modern life favour the worrier. Balancing the worrier-warrior seesaw feels like a uniquely human problem and indeed until recently it was thought that the COMT 158 polymorphism was only found in humans. Recent research though has revealed that the polymorphism exists in other primates.[13] Assam macaques are monkeys that live in large groups with plenty of aggressive social interactions. Studies of their aggression, dominance hierarchies and genetics showed interesting and complex links between which variants of the COMT gene individuals possess and how they express aggression when they are dominant. It isn't quite as simple as 'warrior' versus 'worrier' in macaques, and studies that link COMT to wider characteristics of aggression in humans have found much the same complexity. Differences in susceptibility to certain behaviours and conditions link with ongoing environmental conditions, past environments and genotype to create complex relationships between factors that are not always easy to disentangle from each other. The discovery of a primate with the same polymorphism and similar links to aggression and social stress provides a potential system to disentangle those connections in a much more controlled and experimental way than is currently possible with humans. Of course, the ethical dimensions of such studies are far from straightforward.

Can we 'heal' stress?

If long-term modern-world stress is bad and inherently biochemical in nature, then can we not simply block the biochemical pathways that result in, for example, the production of cortisol? The problem with such a direct approach, as tempting as it sounds, is that our hormonal system is so complex and interconnected that such an intervention would have a range of unwanted effects.

Some indication of this is provided by the suite of symptoms associated with cortisol deficiency, which can occur if the adrenal glands stop functioning correctly (a deficiency known as Addison's disease) or if the pituitary gland is unable to produce the hormones that tell the adrenal glands to produce cortisol. Although cortisol is one of the main mediators between chronic stress and disease, an inability to produce it does not result in better health; far from it. Cortisol deficiency results in fatigue, dizziness, weight loss, muscle loss and mood changes and without treatment in the form of cortisol supplementation it can be life-threatening.

The complexity of our endocrine system and its association (or not) with stress-related conditions does not lend itself to simple 'turn it off' interventions and, as might be expected, the general treatment advice for patients suffering from stress is not straightforward. The NHS in the UK takes a hands-on, problem-solving approach, which I think may be less than helpful if you are stressed. Heading up his '10 Stress Busting Suggestions', Professor Cary Cooper, an occupational health expert from the University of Lancaster, is quoted as saying 'In life, there's always a solution to a problem. Not taking control of the situation and doing nothing will only make your problems worse.' In other words, get a grip and sort out your problems; pull yourself together! I don't disagree with the sentiment, but it also feels a little like shouting to someone dangling off a cliff edge to 'hang on!' It is undoubtedly very good advice, but not especially helpful in the situation. Likewise the top 10 stress busters all sound excellent ('take control', 'help other people', 'try to be positive', 'work smarter not harder', 'be active' and so on) but in reality may be difficult to manage if you are gripped by anxiety and unable to do anything, let alone 'connect with people' or 'challenge yourself'. Other advice on the site includes some breathing exercises, time-management techniques and a recommendation to 'take a holiday'. Having seen some people in the grips of an anxiety episode, advising them to book a holiday seems about as useful as suggesting they 'plan the next Moon mission' in terms of its achievability, at least in the near future. Overall, the advice from a range of sources is neatly summed up by 'It's important to tackle the causes of stress in your life if you can. Avoiding problems rather than facing them can make things worse. But it's not always possible to change a stressful situation. You may need to accept there's nothing you can do about it and refocus your energies elsewhere.' Easier said than done for many, I would suggest.

There is ample evidence to say that our current way of life is bad for our health in a multitude of ways and that the stresses associated with life only seem to be getting worse. Despite the difficulties in disentangling the various different factors that contribute to any particular disease, the links between stress and that disease, and perhaps hardest of all, the different factors that contribute to stress, there is little out there to suggest that stress is getting better. This has likely been the case for some time now, possibly even millennia. The usual pattern is to view human history as a series of great accomplishments, each of which led to longer, safer, more productive lives and a greater measure of what we have come to recognise as 'civilisation'. Thus, we look back at the rise of the modern human through the lens of technological aggrandisement; the Stone Age; the Bronze Age; the Iron Age; the development of agriculture; the industrial revolution; the rise of medicine; the internet and so on. I would argue that, perhaps with the exception of medicine, the same history could be viewed as an ever-increasing opportunity to introduce more stress into our lives. 'Keeping up with the Jones's' probably started the second that we had neighbours and something meaningful to compare with them. I can see no reason to think that early flint knappers for example, pioneers at the very start of our technological progress, didn't experience some sort of stress and anxiety when comparing their efforts with others. Likewise, the ability to accrue 'wealth' in whatever form would lead to stress, whether it be from a modern job or from being an early, successful farmer supplying the local city-state with produce. Where truly modern life excels though, is in the range of potential stresses, their pervasiveness and our inability to shut them down. Modern life is an equal-opportunities stress provider and few seem to escape its ravages in one way or another. Dealing with stress is clearly important but when the best advice available boils down to 'try to refocus your energies', it is clear we are quite a way behind the curve in dealing with the problem medically and socially.

Perhaps the most pivotal technological development in the development of human stress is the rise of the internet, together with the associated rise in mobile technology and resultant 'always on' culture. The evolution of a sharp intellect and well-tuned social skills dramatically intersect with the sudden global rise of the internet in ways that produce some of the most spectacular and damaging mismatches between our evolutionary legacy and the modern world. It is to this uniquely twenty-first-century problem that we now turn our attention.

CHAPTER SIX

Crippling Networks

In the previous chapter I developed the idea that the modern world is full of stresses and that many of these stresses have come about recently. A good proportion of the stresses we met during our 'typical' modern day were focused on something that virtually all of us now take so much for granted that it becomes very difficult to imagine life without it: the mobile phone. Whether it is 'hassle' from social network notifications or the 'always on' culture that phone-based email and enhanced connectivity promotes, it is clear that the mobile, and the virtual world that it enables us to connect with, might be problematic. It is not surprising that we are having trouble coping with this revolution because our phones connect us to the world in ways that even our most recent ancestors (and unless you are a very young reader, I am talking about our parents) could not imagine. Before considering the problems that such technology can cause, and why they stem from an evolutionary history that has adapted us for a very different world, it is worth taking a moment just to realise how powerful those beeping, vibrating, stressful, distracting hunks of plastic, metal, silicon and rare earth minerals really are.

A short history of your phone

If you are aware of any computing in the 1980s you will likely have heard of the Cray-2 supercomputer. In 1985 it was the fastest machine in the world, an absolute beast of a processor used for the sort of applications and organisations that get thriller writers all of a shiver: nuclear weapons research and secret sonar testing for NASA and the United States Government. It was used for other things as well, like designing Ford cars and for less classified research in universities. The Cray-2 was *the* powerful computer of the mid-1980s and yet the iPhone nestling in your pocket is far more powerful. Even the now outdated iPhone 4 had nearly three times the processing power of the mighty Cray-2, and there is absolutely no doubt which device had a better video camera or produced better graphics. Also, with its hefty footprint and need for a complex built-in cooling system the Cray-2 was not exactly pocket-friendly. You might complain about the power consumption of your mobile phone but the Cray-2 needed 200kW, equivalent in car terms to 268 brake horsepower or about the power of a full-spec Land Rover Discovery Sport. In 1969 NASA put Neil Armstrong and Buzz Aldrin on the Moon and guided Michael Collins (in my opinion the most interesting of the three Apollo 11 astronauts) around the Moon to pick them up again using a combined computing power that could be matched by just two Nintendo Entertainment Systems, used by the serious gamer back in the Cray-2's heyday. The latest mobile phones have processing power that is so much greater than NASA in 1969 that it's hardly worth comparing, but if you're interested, the nearly 30kg (66lb) guidance computer onboard the Command Module had a processing speed of 0.044MHz; the iPhone 5 runs at 1.3GHz or about 30,000 times faster. The code required to monitor the status of the flight and the astronauts took up 6MB; the iPhone 5 can store 1GB or about 167 times more, and yet you use it to 'like' a video of a cat sneezing. It's almost insulting…

Powerful processors and massive memory are undeniably impressive but ultimately they are only measures of the potential of phones. It is the penetration of mobile phones into our everyday lives, their ubiquity and, frankly, their absolute necessity that are the true measures of their success. Virtually all of us have them, they are everywhere and they can do pretty much everything. The original function of a telephone was to convey messages using speech, which was a step change in technology from the telegraph, but telephones quickly changed from being conveyors of 'important messages' to instruments for casual conversation. Likewise, mobiles have gone from workaday objects capable of connecting us on the move, the ultimate 1980s accessory for any yuppie with enough arm strength to hoist it to their ear, to the defining technology of our time. They have also become the primary tool that we now use to construct something central to this chapter and central to our modern lives: our virtual social networks.

The rise of social media
Unless you've been living truly off grid for the last 15 years or so, you cannot have escaped the term 'social media'. For a start, many news stories via 'traditional media' these days come adorned with the thoughts and

opinions of a coterie of celebrity Tweeters. Quite often these celebrities are virtually unknown outside of the virtual world in which they are lauded, having become elevated to the status of 'influencer' because of some beautiful Instagram shots attracting thousands of followers or some tweets that happened to 'go viral'. Insta-famous, influencers, tweets, Facebook-friends, 'likes', going viral; literally none of those words and phrases would make any sense to us even 15 years ago. A primary reason for this great rise in social media is the fact that we can interact with the virtual world simply, reliably and cheaply through our smartphones. If we had to log on to a desk-bound computer every time we wanted to update our status or post a photo, uploaded from a digital camera via a lead or card slot, it is hard to see too many of us bothering. Certainly no one would be going through the effort required multiple times an hour. Yet now, because of our phones, for many people the virtual worlds that we now inhabit are as important (or more important) than the very different 'actual world' in which we evolved. Throw in some super-realistic massively multiplayer gaming, the dark web and virtual economics and it's hardly surprising that we're rapidly getting out of our depth.

Before I get too far out of *my* depth, I should define the battleground here. I'm mostly concerned with the aspects of the human animal that have evolved for a world in which we no longer live, or those aspects that were rather useful in our evolutionary past but are coming back to bite us now that we have altered our environment. The science exploring our interaction with the virtual world is still, obviously, very much in its infancy. The psychology of those 'lost' in highly immersive virtual worlds is fascinating, and surely linked with our propensity for abstract thought and our addictive tendencies (Chapter 8). However, it is our relationship with more everyday social networking, and particularly the size of those networks, that turns out to be the aspect best explained within the context of our evolutionary heritage. It is also the aspect of the virtual world that affects most people across the widest demographic because there are 2.38 billion Facebook users and 1 billion active Instagram users (people who use the platform every month). Twitter is fast falling out of favour if social media commentators are to be believed, but even Twitter has more than 320 million active users, roughly the equivalent of the entire population of the USA. Social media provides an ideal context to explore our relationship between our evolved selves and the massively connected world that our technology has created. At least it is for now, but as we've already seen it is a fast-moving world.

At its simplest, social media describes any interactive computer-based application that allows people to create and join networks of other people with shared interests, ideas or need for information, or for pretty much any other reason that people like to connect with each other. Some of the reasons people want to connect are wonderfully wholesome, with networks of people forming on platforms to appreciate the arts, music and books for example. Facebook, the most used of all the social networking platforms, is used by billions across the world to keep in touch with old friends and to form networks of new friends, many of which may never have physically met. There are, of course, many other less wholesome reasons why people want to connect with others and form virtual social networks. There may be a desire to share extreme and possibly outlawed political ideologies, find sexual partners, browse (is that the right verb?) pornography or trade in illicit material and information (child pornography, illegal wildlife products, bomb-making instructions, drugs or guns for example).

I interact on Facebook with a number of people in a manner that would be described as a friendship and yet I have never met them; we are connected through other friends and through shared interests. Likewise, I 'know' a large number of people via Twitter that I have never met in real life. For me, and for many others, these online virtual relationships can be very useful and fulfilling. Professionally, I have developed contacts via social media that have resulted in research collaborations, broadcasting opportunities, speaking engagements and, now that I think about it, even put me in touch with the commissioner of this book. Social media can be, and often is, a very important and valued component of our lives. I can honestly say that Twitter and Facebook have enriched my life and I enjoy using them. However, judging from the number of people I now see posting that they are 'giving up' Facebook, or 'taking a break' from Twitter, it is clear that not everyone feels the same.

Is social media harmful?

The idea that social media and online social networking might be harmful to us in some way began to form almost as soon as social media started to take hold. In 2006 the social media landscape was very different to the one that is familiar to us now. Facebook wasn't launched publicly until September of that year, Twitter had only launched in July 2006 and Instagram was still a digital twinkle in the eyes of Kevin Systrom and Mike Krieger, who launched the company four years later. Before the Big Three came to dominate in later years, two of the biggest players were Friendster and Myspace. Friendster was a social networking site launched in 2002 that was rapidly eclipsed by Facebook, morphing into a social gaming site in 2011 before being suspended in 2015 and then finally being put out of its misery in 2018. I'll be honest, the first I'd heard of it was in reading around for this section. Myspace, although largely unknown to younger users now, was the largest social media platform in the world between 2005 and 2008. Unlike Friendster, I had heard of Myspace and indeed once had a Myspace account. Like Friendster though, it proved to be another victim of the relentless rise of Facebook, although it does still exist, at least at the time of writing. Stylised now as 'myspace', it has suffered from a string of setbacks in recent years including a major data breach and the loss of 50 million songs and 12 years of uploaded user content.[1]

Despite their lack of long-term success, back in 2006 both Myspace and Friendster were helping to define social media as we know it and their early success in the sphere was attracting interest from social scientists keen to explore the wider consequences of online social networking. Researchers from the Netherlands studied adolescents using Friendster and Myspace, and they identified two potentially troubling aspects of online social networking that have continued to be a cause for concern over the decade and a half that followed.[2] Specifically, they examined well-being and social self-esteem based on feedback that the adolescents (aged 10 to 19) using the social networking sites received from their online friends. What they found out seems now to be obvious. Positive feedback on their profiles (which these days we would refer to as positive comments, likes, shares, retweets or increased followers depending on the platform used) enhanced users' well-being and self-esteem while negative feedback (negative comments or a lack of interaction) had the reverse effect. Both positive and negative effects were increased with greater use of social networking sites. Even then, in the early days of social media, users were riding the wave of positive feedback, building up self-esteem and well-being through the online validation heaped on them by their virtual network, but also spiralling down in a 'crash' of negative feedback. As a Twitter user, I will admit freely that having posts that stack up likes and retweets, getting favourable comments and attracting the attention of 'super users' with millions of followers is a bit of a high. It is certainly addictive and I can well imagine how the reverse would also be true, although luckily I am adept at ignoring or 'muting' those who deal in negativity. Others are not so fortunate, as is easy to see on any social platform where people have been caught up in a maelstrom of hate and scorn.

It is interesting to see that early investigations into the potential for social networks to cause harm examined self-esteem and overall well-being. These aspects have not gone away in the last decade and a half, with self-esteem worries exacerbated even further by ever more glamorous status updates on Facebook and the rise of Instagram. Instagram is essentially a photo-sharing platform, but the easy availability of filters that can turn even an average holiday snap into something far more attractive consi-derably elevates it above other photo-sharing sites like Flickr. When all is said and done, Instagram is really about posting photos that say 'look at how amazing my life is' and the 'Insta-fabulous' lifestyle is immediately clear to anyone who joins the platform. Intuitively, conscious and subconscious comparison of the seeming tedium of one's own day-to-day life with the heavily filtered highlights of everyone else's makes blows to self-esteem seem inevitable, even for the sturdiest of egos. For once, the evidence supports our intuition. Surveys of social media users report that social media sites make more than half of users feel inadequate or unattractive. Meanwhile, studies of selfie viewing (people looking at images of themselves that they have themselves taken) found that it lowered self-esteem. Frequent selfie viewing was actually able to lower users' life satisfaction, especially for individuals with a high need for popularity. Interestingly, 'groupie' viewing, looking at images of groups of which you are a part and are being taken by someone else in the group, had the opposite effect; it increased self-esteem and life satisfaction.[3] Selfies encourage direct comparison with others in what is termed 'upward social comparison', a tendency to compare yourself with people whom you perceive as being better. Groupies on the other hand don't encourage this comparison. Instead, presumably viewers focus more on the positive feeling associated with being part of that group (usually of friends) rather than the negative feeling of not being someone else.

Further studies on the negative effects of social networking and social media fleshed out the same story and conventional media, threatened perhaps by what might have appeared to them at the time as an existential crisis, were all too keen to fan the flames. Stress of course featured heavily, along with its bedfellow, anxiety (Chapter 5). Add into the mix depression, addiction (more of which in Chapter 8), sleep deprivation (which of course leads to stress and anxiety), relationship crises (there's some more stress for you), envy and loneliness, and it would seem most of modern life's ills have at some point in recent years been blamed on social networking. The evidence supporting some of these associations is, you will not be surprised by now to learn, complex. Not only can it be hard to quantify reliably some of the effects, but also monitoring and standardising social media use and controlling for the effects of underlying conditions mean that the sort of broad-brush conclusions that we like are very hard to reach.

Depression is perhaps the most complex of the purported negative associations with social networking. Certainly, some studies have found links. In 2012, for example, two studies investigated the link between depression and social media use in more than 1,700 subjects with an average age of around 19–20, rather endearingly referred to as 'youths' by the researchers. The questions we can ask can be dressed up in all kinds of fancy sociological and psychological clothes but fundamentally what we want to know is, does social media use cause depression? The researchers concluded that, with regard to frequency of social networking, the answer to this question is no; they found no evidence that more frequent usage is associated with depressive symptoms. This is superficially reassuring, but the researchers probed a little deeper and were able to reach a more nuanced and frankly more disturbing conclusion. Negative interactions while social networking were associated with increases in depressive symptoms over time, building up feelings of worthlessness, hopelessness and low mood cumulatively with use. They frame social networking as 'a salient venue in which youths experience the depressogenic effects of poor quality relationships'; in other words, yet another place to get dumped on by the world.[4] When this finding is combined with another from the same research, some tentative evidence that some young people may be more at risk than others, then alarm bells really start ringing. Young people prone to 'depressive rumination', focusing on symptoms of distress and their possible consequences leading to an unhealthy fixation on negative feelings and problems, may be more likely to experience depressive symptoms as a result of negative interactions on social media. Such individuals may also tend to feel more depressed following social networking interactions. So, while there may not be a clear-cut relationship between social networking and depression, there are clear signs that for some of us social networking can be a contributing factor in degrading mental health. Perhaps worth thinking about, the next time you decide to make a comment on some stranger's post?

Another study, in 2016, examined the links between depression and anxiety in 1,700 people who used multiple social media platforms. Their results were very clear: people who used multiple platforms were at a threefold risk of depression and anxiety. This shocking headline figure, however, is actually a result of comparing people who use between seven and eleven platforms, which is frankly an awful lot, compared to those who use between zero and two, a number that might be considered more normal. Some of the reasons suggested to account for this increase in depression include having a distorted view of other people's lives, feeling like time spent on social media is a waste and cyber-bullying. Cyber-bullying, which is not likely to be going away any time soon, is an area where links between social networking and depression, and the role of depressive rumination,[5] have been clearly seen in young people. This should be of great concern given that young people have taken their own lives as a consequence. Young people killing themselves (perhaps the ultimate maladaptive human behaviour) as a consequence of interactions in a virtual world of social networks would, 20 years ago, have been the plot line of a dystopian future novel. Now, it is an understandable and serious concern of many parents as their children navigate the world we have created. It's worth saying this again: we have created a virtual world where young people are killing themselves (and, to take an evolutionary angle, leaving no offspring) because of interactions in that artificial world.

Aside from depression, there is evidence linking anxiety in general to social media use, although there are also studies suggesting that social media can reduce anxiety.[6] Social media and mood seems more straight-forward, at least according to one study that concluded that Facebook activity negatively correlates with mood; the more you use it the worse you feel. They did not find the same effect with internet browsing, so it isn't simply a case that being immersed in the online world or screen-time per se are 'bad', rather it is something about the social networks of Facebook that causes low mood. They concluded that the reason for this is that people feel bad because they have wasted their time on Facebook. Interestingly, the negative effects of use don't stop users making what is described as a 'forecasting error', with people expecting to feel better after engaging with Facebook despite experiences clearly to the contrary.

Anxiety, stress and depression can all lead to poor sleep, and poor sleep can cause stress and anxiety. Throw social networks into this toxic nocturnal feedback loop and you have another medical issue: insomnia. A study of 18–30-year-olds found a link between social network use and sleep disturbances, with the conclusion that the blue light emitted by smartphones had a part to play, but how often participants logged on to social media sites, rather than the time they spent on them, was a better predictor of disturbed sleep. The results suggest obsessive checking of social media is a problem, but they couldn't determine whether social media causes disturbed sleep or whether people with disturbed sleep spend more time on social media.[7] Cause-and-effect is a harsh mistress.

It seems fair to say that although the picture is complex, there are some links and relationships between specific disorders and conditions, like depression, anxiety and insomnia, and social network use. But what about more general and non-medical measures of well-being or happiness? Does social networking make us sad? Well, probably overall the answer is no, at least according to one study in the Netherlands published in 2018. The study used data from the Dutch Longitudinal Internet Studies for the Social Sciences (LISS) panel for the years 2012–2013. In this survey, participants reported on numerous aspects of their life including their happiness, social networking site use and a variety of aspects of their lives that enabled researchers to infer the quality and quantity of contact with friends and family. It is this last aspect of the study that threw up something interesting. The study revealed a negative association between the total time spend on social networking sites and happiness for users of those sites who were socially disconnected and lonely. This result held even when the researchers allowed for other factors, like hours spent on other internet sites or household income.[8] Once again the pattern seems to be that some people are especially prone to negative experiences associated with online social networking.

The paradox that emerges from the findings of many online social networking studies is that, consistently, research shows that friends make us happier, healthier and lead us to live longer. Our friendships are our real-world social networks and having strong social networks is universally regarded as a 'good thing', supported by robust empirical evidence.[9] Why then might online social networks, virtual friend groups, be a bad thing? We have already seen some of the reasons put forward to explain this negative association, such as a sense that being on Facebook is a waste of time, but these are not connected with any particular mismatch between the contemporary world and our evolution, at least not in a direct or meaningful way. However, there is an aspect of

our evolutionary past that does create a mismatch with
the online world, and that might go some way to explain
some of the issues we have with the ever-increasing
sprawl of our virtual social networks. In a nutshell, our
brains are too small to cope. To find out more we need
to visit the zoo.

It's all about grooming

The next time you're at the zoo take a wander over to
the primate section and spend some quality time looking
at our nearest biological relatives. I don't just mean a
cursory couple of minutes noting that some of them
have weird backsides, although the 'ischial callosities' on
the buttocks of baboons, macaques and chimpanzees are
hard to ignore. I mean actually watch their behaviour.
It won't be long at all until you spot them grooming.

Self-grooming, keeping yourself clean and free of
parasites by using the mouth in combination with
whatever appendages evolution has provided, is a
common behaviour throughout the animal world. As
the song goes, birds do it, bees do it and even educated
fleas do it. Invertebrates (like bees and fleas) groom to
remove dirt, fungal spores, mites and other potentially
undesirable hitchhikers from their exoskeleton. Birds
and mammals on the other hand have a more complex
external surface consisting of skin, which needs clearing
of parasites, and further external structures that function
best when clean and properly set up. Feathers on birds
need to be 'zippered up' to provide a better flying surface
and fur on mammals needs to be clear of substances that
cause matting and a loss of insulation. In addition to
removing unwanted passengers like mites and lice,

the sometimes complex contortions involved in self-grooming (or preening as it is known in birds) can be a way to distribute oils over the feathers and fur. Birds have a uropygial gland located on the top side of the base of their tail that secretes a substance called preen oil. This oil helps to waterproof and condition feathers, making birds better insulated and able to fly. Mammals, including us, have sebaceous glands that secrete oil that lubricates and waterproofs the skin and hair. Self-grooming, then, is a very common behaviour and once you get your eye in for the tell-tale movements you will start spotting it regularly, whether you are watching spiders or sparrows, butterflies or bears. What is more uncommon in general in animals, but will be very clearly apparent in many of the primates you see at the zoo, is the phenomenon of social grooming: grooming and being groomed by someone else.

Social grooming in primates has benefits that extend well beyond the fact that someone else can reach the parts that you can't. Certainly it is helpful to have another pair of hands to remove those especially stubborn parasites, clean off that ground-in dirt and untangle those proto-dreadlocks, but social grooming has more subtle effects too. One effect is that grooming can reduce heart rate, calming down both groomed and groomer. This observation (in macaques and likely to be the case in other species too) is surely related to the now well-supported fact that stroking cats can reduce heart rate and blood pressure in humans. Indeed, 'socially grooming' your cat has been shown to reduce the chance of stroke, a major killer related in many cases directly to high blood pressure, by up to a third. Similar effects of this value of 'inter-specific' grooming, whereby one species grooms another, comes from studies of rhesus monkeys. Familiar humans that groom rhesus monkeys produced the same calming effect of reducing heart rate as was produced when individuals were groomed by another rhesus monkey. It may be the case that the calming benefits of grooming don't even require another living being in order to be expressed. I once watched a captive colobus monkey apparently grooming a soft toy, devoting attention to the furry rabbit with a look of blissful tranquillity on its face. This was especially notable, since the groomer was a male in a group of males, the majority of which were directing most of their 'grooming attention' to their own genitals.

The fact that our closest relatives, the chimpanzee and bonobo, spend 10 per cent or more of their time socially grooming[10] and that we have a physiological response to grooming-like activities strongly suggests that we have an important evolutionary history when it comes to social grooming. Watching humans interacting reveals many behaviours that would be considered social grooming in any other primates, from brushing each other's hair (although as the father of three daughters with long hair I do not consider this to be a very calming interaction) to picking lint off each other's clothes.

But where does social grooming fit in terms of considering how evolution has left us unprepared for the modern world? One clue to this comes in the broader use of the term 'social grooming' within behavioural science. Social grooming is often used to describe not just those behaviours associated with maintaining the condition and quality of skin and fur but also behaviours that facilitate the construction of social relationships.[11] Social grooming is all about connecting with individuals in ways that develop and nurture alliances and coalitions, cementing and securing relationships and friendships that provide benefits to those involved in the grooming partnerships. In fact, the consensus across most studies of social grooming in primates is that its most critical function is not in removing parasites or improving fur condition (which may have been the initial selective advantage in its evolution) but in establishing, developing and reinforcing social networks and relationships. To use the sort of language that most primate researchers assiduously avoid, when monkeys are picking at each other, unmatting fur and eating ticks, what they are really doing is making friends. By developing good social networks through the act of grooming, primate groups can become more stable, which could bring benefits to individual members through strength of numbers and competitive advantage. It is better for the individual to be in a better group, and a better group is one that has strong and stable relationships between group members. It also forges alliances within the group, and in a number of species scientists have observed monkeys backing up other monkeys in confrontations in ways that really do seem best described as friendships. These coalitions are closely related to grooming relationships, and grooming can be linked to a great many of the different facets of a group's dynamics. By understanding the benefits of social grooming from a social angle we can develop just the sort of perspective we need to understand the differences and similarities between picking a hair from a friend's face and friending them on Facebook.

The power of gossip

Social grooming in its wider context includes behaviours that are not connected to the physical act of grooming. Gossip, for example, casual conversation between people about people, is considered to be a form of social grooming and like other forms of grooming it has a beneficial physiological effect. Studies have revealed that gossiping, as opposed to other forms of conversation, reduces cortisol (and so reduces stress – see Chapter 5 for why that is such a good thing) and increases the level of oxytocin. Oxytocin is a hormone that plays an important role in social bonding, sexual reproduction, childbirth and the period after childbirth, when it is associated with maternal bonding and the onset of lactation (we will meet it again when we consider trust in Chapter 9). Gossiping has the twin benefits of reducing stress and bringing people together, acting to bond individuals in just the same way that more obvious physical grooming can achieve.[12] Such is the importance of gossip in human behaviour that it has been hypothesised to have played a part in the evolution of language. Robin Dunbar, an anthropologist, evolutionary psychologist and primate behavioural expert, proposed in 1996 that language in humans evolved as a response to increasing group size. The logic behind this hypothesis is that conventional social grooming becomes far too inefficient in larger groups; there are simply too many other backs to scratch and not enough of your back that's itchy. The benefits to individuals of being in larger groups (being able to defend yourself more effectively, to bring down larger prey and so on) would have made the building of stable alliances important, but the need for 'manual' grooming to create and maintain these social networks was simply too time-inefficient to get much else done. The beginnings of language, Dunbar proposed, were a form of cheap and efficient 'vocal' grooming, starting with the sort of mutually pleasing grunts that allowed groups to bond more efficiently. Over time, vocal grooming sounds evolved into language, with a myriad of additional advantages shaping its subsequent development.

Dunbar's gossip-grooming hypothesis has been criticised. First among these criticisms is the fact that manual grooming takes time and effort and therefore provides an honest and reliable signal of friendship that simply cannot be matched by vocal grooming; in other words, talk is cheap. Criticism notwithstanding, the Dunbar language hypothesis has at its heart a feature that is of great interest in unravelling the potential links between our evolutionary heritage and online social networking. Group size, Dunbar argued, was a critical factor in the evolution of language because at some point in our evolutionary history, group size prevented conventional social grooming from playing its crucial role in developing and maintaining real-world social networks.[13] Group size is also critical to Dunbar's most famous work, and the idea that now bears his name: the Dunbar number.

The Dunbar number
The Dunbar number is a measure of our social limitations. It is the maximum number of people with whom an individual can maintain stable social relationships in which individuals within the group know who each person is and how each member of the group relates to

the others. It is a measure of our maximum social network size and for humans it works out somewhere between 100 and 250, often stated as 150. Dunbar's number is basically the number of people you can know and keep social contact with. It isn't the total number of people you might 'know of', and neither does it include people you may have once been close to but no longer associate with. Dunbar described it as 'the number of people you would not feel embarrassed about joining uninvited for a drink if you happened to bump into them in a bar'. Dunbar's number is a measure of the number of people in a social network that you can feel 'comfortable' with; the sort of network that provides the kinds of benefits that we see from having friends. It's our social feel-good number.

Dunbar didn't simply magic this number from thin air. He came to this conclusion by considering us biologically, as a primate, and comparing us with other members of our biological order.[14] What Dunbar found was that group size in primates related nicely to the volume of the neocortex of the brain. The neocortex is that part of the mammalian brain involved with 'higher-order' functions like thinking, spatial reasoning and language. Dunbar suggested that the reason behind the relationship between group size and neocortex volume across 38 primate genera was that the number of relationships an individual can simultaneously monitor is limited by the information-processing capacity of the brain. This processing capacity is itself limited by the number of neurons (measured using the proxy of volume) of the neocortex, leading us to the obvious conclusion of bigger brain, bigger network. When the

maximum network size (which subsequently became
known as the Dunbar number) is breached, the group
can become unstable and begin to fragment.

By generating a predictive relationship between
neocortex size and group size it is possible to work out
the group size for humans based on our brain size, and it
is this approach that leads us to a human Dunbar number
of around 150. Dunbar went on to develop and test this
idea and found considerable support for it from a range
of sources, including group sizes in Neolithic farming
villages, Roman army unit sizes and even career-seeking
via social networks.[15] It is an idea that is intuitively
seductive and seems to tie in with our experiences. If it
is true, then it would certainly go a long way to account
for some of the issues we might have with online social
networking, where we have near-constant interactions
with networks of individuals orders of magnitude greater
than our evolved social limit. Indeed, the Dunbar number
is a widely reported and widely applied measure of our
social ability. As the famous American biologist Paul
Ehrlich said in an ABC broadcast relating to religion in
2015, 'we are a small-group animal' looking to survive in
a connected world of billions. When asked to elaborate
on this by The Conversation Fact Check* he added 'The
Dunbar number is ~150, size of hunter-gatherer groups,
length of Christmas lists, and so on. My point was we're
a small-group social animal now suddenly (in cultural

* The Conversation Fact Check is a part of The Conversation website.
Specialists analyse stories from a wide range of topics including
politics, science and medicine to get to the facts. As they say, they
'remove the spin from the debate'. https://theconversation.com/
uk/factcheck.

evolution time) trying to find ways to live in gigantic groups.' He didn't feel the need to justify the number further or to provide the usual cautionary qualifiers that litter most scientists' public utterances. The Dunbar number seems then to be widely accepted; but does it, and its implications for modern life, stand up to scrutiny?

Digging deeper into the Dunbar number

As a descriptive and explanatory tool in the social sciences and humanities, the Dunbar number has been widely supported and influential both in theory and practice. For example, in his excellent book *The Tipping Point* Canadian journalist and author Malcolm Gladwell describes how the company W.L. Gore and Associates, best known for the Gore-Tex brand of waterproof breathable fabric, runs its business according to the Dunbar number. Its manufacturing is organised in such a way that each product can be accommodated in units employing no more than 150 people, with everyone in that unit being housed together.[16] Sizing and organising according to the Dunbar number creates an environment where everyone knows everyone else and their relationships to others in the group, and is available to each other. If the group size exceeds 150 then divisions are initiated to create new 'Dunbar optimised' groups. For this company at least the approach has been very successful. In other fields though, the entire foundations on which the Dunbar number is based have come under attack.

The Dunbar number rests entirely on the correlation between neocortex size and group size in other primates. One criticism of this approach is that a great many factors come into play when considering the evolution of the brain and especially the neocortex. Humans are pretty big, and large animals tend to have a large neocortex; a sperm whale neocortex is nearly 10 per cent larger than ours in proportion to its body size. Other factors like territory size, diet and activity patterns through the day have also been found to explain neocortex size just as successfully as group size, and overall the correlations and patterns are complex. A critical review of comparative studies of brain size found 'a number of substantial problems' with assumptions, data collection and subsequent analysis that were 'particularly apparent in those analyses in which attempts are made to correlate complex behaviour with parts of the brain that carry out multiple functions', which is the neocortex in a nutshell.[17] Overall, the general mood seems to be that while social behaviour and neocortex volume may well show a correlation, correlations do not imply causality and this is very often the case when complex, interrelated and highly entangled traits are being considered.

Further cracks seem to appear when the anthropological evidence is considered. The anthropologist Frank Marlowe has suggested that hunter–gatherers with overnight camp groups, or bands, form into units of 30–50, group sizes that do not support Dunbar's ideas. Dunbar has argued against this, pointing out that these bands are often unstable and are embedded in larger local communities.[18] Comparison of human groups from hunter-gatherer bands to modern companies actually leads to an important and currently unanswered question: what sort of human social unit is appropriate to study if we want to draw inferences about human social evolution? Despite these criticisms, Dunbar asserts that

'There is now considerable evidence that groupings of this size occur frequently in human social organisations, and that this is the normative limit on the size of personal social networks among adults.' He is right about the evidence, but he was also a co-author on many of the studies he cites in support. That's not a problem per se, but it would be good to have something even more solid on which to hang one's scientific hat. Luckily, away from the vagaries and difficulties of social anthropology and the debates over the appropriate groups for meaningful study, there is now some solid neurological evidence supporting Dunbar. A number of studies have reported correlations between individual differences in social network size and the volumes of areas in the brain involved with social cognition, including the cortex and, intriguingly given its connections to stress (see Chapter 5), the amygdala.[18]

Dunbar's number isn't the only number in town. Some field studies in the United States led anthropologist H. Russell Bernard, Peter Killworth and colleagues to a mean average number of 290 (roughly double Dunbar's) and a median for the now so-called Bernard-Killworth number of 231. The median is the 'middle value' in a distribution of data and the fact that this is lower than the mean indicates that some individuals had very large networks, pushing the mean average up. Although far less well known, the Bernard-Killworth number is nonetheless quite well supported using different methods, including asking people about their social networks and using statistical approaches to estimate people's networks.[19]

Overall, despite some criticisms and a competitor in the shape of the Bernard-Killworth number, Dunbar's number has pretty much survived the test of time. The details may have come under fire, but the notion that there is an upper limit on our ability to keep track of meaningful social interactions is now widely supported. What is more, set against our potentially vast online social networks, the exact value of this number is actually not that important as long as it remains, as it seems to be doing, in the range of a few hundred or so. Who cares if our upper limit is 150, 290 or even 500 when many of us are juggling active social networks that might well be in the 1,000 to 10,000 range?

We can get further insight into possible evolutionary constraints on our social networking through more detailed analyses that show that the relationships within the social network encapsulated by the Dunbar number are not the same. There are layers within the network that occur around 3 to 5, 9 to 15 and 30 to 45 individuals, with each successively larger layer incorporating individuals with progressively weaker ties, expressed both as perceived emotional closeness and frequency of interaction.[20] These layers might explain the consistent and seemingly spontaneous formation of smaller social groups around these size bands, and the techniques used to identify these preferred group sizes can also account for layers beyond the Dunbar number, at 500 and 1,500 people. In these outer reaches of our social solar system, at 1,500 people, we reach the limit not of our cognitive ability to keep track of interactions but of something much more basic: our ability to remember that many faces.[21]

This layered analysis of social networks is neat and, like the overall theory, highly intuitive, but once again it can start to crumble around the edges when placed under pressure. Studies have shown that relationships simply aren't tiered like an emotionally simplistic layer cake and that human social interactions are more complex and interconnected. Different relationships within the layers provide support for different things; thus a close family member might provide regular emotional support, while a friend in the next layer might provide occasional but exceptionally important financial support. In practice, there are multiple ways to define the 'closeness' that leads to people being placed in different layers.[22]

Studies have also attempted to define far larger networks than those estimated by the Dunbar number. One study in 2006 found that Americans know a mean of 610 and a median of 550 people. The range in network size was vast, with 90 per cent of the adult population knowing between 250 and 1,710 people, and half knowing 400 to 800 people. Another study, which followed people around in Malta, showed that people had networks of people that embed them in society, open up opportunities, provide useful services and give a sense of belonging that number around 1,000. To use these studies as arguments against the Dunbar number (as some have done[22]) is perhaps a little obtuse since Dunbar was not including these sorts of relationships in his definition, but overall it matters not. Whichever way we slice it, and despite some heated opposition, the conclusions actually turn out to be more or less the same: we have a limit in terms of social interactions that is imposed by the limitations of our brain, and that limit is likely in the order of hundreds rather than thousands of people. Even if we take the upper limits of those studies that do not support the

Dunbar number we still max out around the 1,500 mark, which is a long way below the online social networks of a great many people, me included.

Is social media a mismatch?

If we accept that we have a constraint on the size of social networks in real life imposed by evolution, and that those constraints conspire to max us out at some level below what many people would consider to be their online network, then can we attribute any blame for the problems caused by social media to the mismatch between the evolved real and the created virtual world? I think the answer is a cautious yes.

The argument made by Dunbar in his 2012 analysis of social networking sites is that online networks were actually not especially large. Bearing in mind the date of the study, Twitter communities were found to be of the order of 100 to 200 (slap bang in Dunbar number territory) and 'Facebook's own data suggest that the claim that large numbers of friends is the norm is, at best, an exaggeration: while the distribution is certainly "fat-tailed" [some people have a lot of friends], the average and modal [the most common value] number of friends is in fact approximately 130'.[18] What is interesting is that in 2019 the number of Facebook friends has only gone up to 155 and once again we find ourselves right in the Dunbar comfort zone. Twitter has slipped out of the zone though, with 707 followers on average in 2019, but even this value is still within the natural network range suggested by other researchers. Dunbar's interpretation of larger group sizes on social media is either that such users are professionals using social media to reach audiences or that people are dragging 'acquaintances' from the 150 to 500 layer into their 'friends' group, in effect promoting people to friends but probably not interacting with them too much and certainly not feeling comfortable to sit down in a pub with them uninvited. The quality of those interactions is also often quite low-rent; a quick 'like' takes much less effort than a supportive conversation in real life.

The average size of online networks does not seem to indicate that we are in any sort of mismatch but such raw figures, useful though they are, do not tell the full story. For one thing, many users (the proportion of which will depend on the distribution of group sizes across all users) will have considerably more than the average-sized group and find themselves well outside of the comfort zone afforded even by the top end of normal social-group size estimates. Social networks online also exist separately from real-world networks, although with some overlap. Consequently, you might be close to your Dunbar number (or whatever estimate you choose to use) in real life and then easily max it out by having even a small network in the virtual world. We evolved for one world, and the Dunbar number and other models of social interaction do not really allow for two networks running side by side in different worlds with perhaps only limited crossover. I would argue that a great many people likely have a combined, complex social network that integrates real and virtual worlds (with some potentially difficult connections between them), the magnitude of which puts them well in excess of the social networks we have cognitively evolved to manage.

A simple count of friends or followers also greatly underestimates the connectivity and sheer neediness of online social networks. Depending on your privacy settings it is possible, indeed likely, for complete strangers to intrude on 'your' network, leave comments, butt in on conversations, challenge your opinions, insult you and even threaten you. These interactions, arising as a direct result of your social networking, would simply not be tolerated in the real world, but we are all still learning and creating the rules of engagement for the virtual world. Again, depending on your settings, each of these intrusions is a needy little notification requiring our attention, distracting us from activities and possibly our real-world network. Also, as we have seen, if such interactions are negative they can add greatly to the potential for anxiety, stress and depression, at least in some of us.

We have evolved to formulate and support small social networks that provide support and friendship and through that a series of selective advantages. We are a social species and we are stronger together. Over time we undoubtedly developed larger and larger social networks, but until recently such networks were not implicated in clinical conditions like depression or seen as sources of stress. Indeed, there is a strong body of research showing that friends and social interactions are overwhelmingly positive. In the last decade or so, though, we have created an online world that has become for some of us a major component of our real-world environment and which is perfectly suited for fostering very large, and potentially massive, networks. These networks may not require the level of individual attention that real-world friendships require, but collectively the scale of interactions that can be entered into, even with relatively modest social media networks spread across just a few platforms, can rapidly become an overriding obsession. Our brains simply aren't allowing us to keep up. My advice: if it's not working for you then don't engage with it. Use it for what you want, and make sure you understand privacy and notification settings, because a constantly beeping mobile phone was definitely not part of our evolutionary background. Actually, that might make a good tweet; maybe it will go viral...

CHAPTER SEVEN

An Unusually Violent Species

Violence is everywhere. Turn on the TV, even during the daytime, and you can see Magnum PI and the A-Team kicking ass as you eat your lunch. Factor in streaming services and you can watch the latest ultra-violence, far more gruesome than the over-choreographed punch-ups found in 1980s reruns, 24/7. Go to the cinema and you'll see that many blockbuster movies come with a heavy side-order of cracked heads and bloodshed. Perhaps the entertainment industry, with what seems like an obsession with extreme violence, might just be mirroring real life, at least if you can believe the headlines? However you choose to 'consume' it, the news is an endless churn of ever more sickening examples of human-on-human violence directed against women, men, children, disabled people, minorities, pensioners; indeed no sector of society seems safe against the onslaught. We punch, kick, bottle, club, stab and shoot, we injure, maim, torture and kill, and we do it for revenge, to rob, to hurt, in anger, premeditated or without any particular thought at all. All of that violence is before we factor in the horrors of war, where political motivations, religion, fear, greed and frankly who the hell knows what create death tolls so large they become the sort of horrifying statistics that magnitude and familiarity have rendered mundane. Assault, rape, bodily harm, manslaughter, murder, crimes against humanity and even genocide dominate the headlines, while millions pay to watch organised, rules-based violence in the many fighting sports that we have devised to vent fury and raise revenue. With all the violence that surrounds us, you might be forgiven for jumping to the obvious conclusion that mankind (and here it really is appropriate to say *man*kind) is an inherently violent species, genetically predetermined to bust heads and spread suffering wherever we go.

Before we dive too deeply into the darker side of human nature (and of course, nurture), let's define our terms. More accurately, let's let the World Health Organisation define them for us: violence is 'the intentional use of physical force or power, threatened or actual, against oneself, another person, or against a group or community, that either results in or has a high likelihood of resulting in injury, death, psychological harm, maldevelopment or deprivation.' You might think that this is rather obvious and that violence falls nicely into the 'we know it when we see it' camp, but the literature on human violence is littered with words with everyday meanings that have more nuance in a scientific context. Violence, for example, is seen as being different from aggression, which was defined back in the 1990s as behaviour 'produced to cause physical harm or humiliation to another person who wishes to avoid it'. On the face of it, quite similar, but recent work has considerably refined this definition to include explicitly the notion of evolutionary advantage, framing aggression as behaviour 'intended to increase one's own dominance and, thus, reproductive success', a definition that can include a great many behaviours we would not regard as 'violence'. Aggressive behaviours, at least according to some commentators on the topic, can include standing up for one's beliefs, being assertive, defending others in need, pursuing careers in law enforcement and the military, business and legal affairs, sports, politics and even scientific debate. Violence, then, is a subset of aggression.[1] A further subset of aggression, and indeed a subset of violence, is lethal violence. Killing another human represents the extreme end of the human aggression spectrum, and because it is arguably the type of behaviour that is of most concern to twenty-first-century society it will be the focus of this chapter. The role of aggression and the assertion of dominance in everyday life can be clearly seen, from the often toxic political debates playing out currently to the rapacious nature of big business. If the tendency towards such behaviour is the result of evolution, and I think it would be hard to argue that natural selection and genetics won't have played some part, then these behaviours fit just as snugly here. But it's my chapter and I'm going to focus on lethal violence, and if you don't like it we can have that 'conversation' outside...

Are we evolved to be violent?

Against a backdrop of apparently ubiquitous human violence, and a history that is literally 'one bloody thing after another', it is all too easy to piece together an evolutionary scenario that leads us to the inevitable conclusion that violence today is a mismatch between modern society and our evolutionary past. Such a scenario could run very much as follows (and yes there are plenty of potential problems here that we will get to shortly).

Violence is, at times, of great benefit to violent individuals, regardless of the species. Males that can out-compete other males by being stronger, by being more violent, will get increased access to females and therefore will get more offspring. Regardless of sex, a 'no nonsense' approach to sorting out competitors will give increased access to territory, food, water and other resources like nesting material. Violence can solve some pretty pressing problems.

We are a social species. Early (and indeed not so early) humans might have profited from being violent towards other groups, outcompeting them for resources and benefiting members of the winning group. Such groups are often family-based, providing a background of shared genetics. Violent individuals may also have gained from being better able to defend themselves personally from attack.

Violence could be a component of more complex behaviour, governed perhaps by testosterone levels, which might be linked with other potentially useful traits like 'ruthlessness' or 'assertiveness'. These traits gained violent individuals further social advantages within their group. It is all too easy to think of scenarios where violence could be directed towards members of the group and towards even close kin, and that such violence could work out as an advantage, perhaps disposing of individuals seeking to take over your status. For homework, watch *Game of Thrones*, although be mindful of the idea that emotions like guilt have been proposed to have evolved to curtail such behaviour (Chapter 5).

Individuals that were violent could have gained status in social groups as a consequence of that violence (again,

Game of Thrones will get you into the right way of thinking here), and thereby gained more offspring, as long as we assume that status equals sexual opportunity. And let's be honest, that is often the case.

If there was a genetic basis to violent behaviour, then any advantage that translated into more offspring could lead to the evolution of violence as part of the complex suite of behaviours associated with social behaviour. Ergo, we evolved to be a violent species. A huge benefit in some past environments, this is a real mismatch in the modern world where, despite what Hollywood might have us believe, going around beating people up is pretty much always a big disadvantage.

The basic logic falls out nicely and resolves down to a simple idea. Violence was useful in the past and that led to the evolution of violence that doesn't suit the modern world; bingo, we have our mismatch. Armchair evolution isn't going to wash here, though, and to work out whether there is any real truth in this Just So story will take us to some difficult places. First, we need to answer the most basic question: are we actually a violent species or are we judging ourselves too harshly?

It would be lovely to be able to say that humans are not in fact a violent species; that the concentrating effect of 24-hour rolling news linked with a certain morbid fascination with gruesome details leads to the perception of violence in society being mismatched with the reality. Sadly, we cannot say this or indeed anything close to it. What we can say is that, compared to other mammals, we are exceedingly violent. Seven times more violent if we take lethal violence as our measure, and that does feel like a pretty good measure.

Before examining the damning evidence of our violent tendencies and how we arrive at a figure of 'seven times more violent', we need to consider violence in the round and from a zoological rather than sociological perspective. Violence within a species, conspecific violence, is not limited to humans as even a cursory examination of the animal world will confirm. Watch ants as they encounter ants from another colony; the workers (all female, by the way) can rapidly escalate from aggressive curiosity with their antennae to full-on biting and shoving, often resulting in decapitation, dismemberment and death. Take a look at some male blackbirds at the height of their breeding pomp if, despite some vocal singing to show where territories start and end, one bird decides to overstep the mark. The resulting tussles are aggressive and decisive although not, as far as I am aware, fatal. I once had a very pleasant evening sundowner disturbed by the sounds and sights of ferocious fighting between two adult male white rhinos having a very serious, and pretty violent, set-to. Take a look at a lion's face or the flanks of an old male hippo and the scars they bear clearly tell the tale of conspecific violence. Violence between individuals within a species is so common within the animal kingdom because at some point animals are competing for resources that are in short supply. There may be all kinds of fancy displays that have evolved to keep violence in check, because even if you win you might get hurt, but when push comes literally to shove animals are more than ready to roll up their metaphorical sleeves and get stuck in.

I have often heard it said that humans are the only species capable of lethal conspecific (within a species) violence. This is incorrect and hopelessly, if rather endearingly, naive. Lethal conspecific violence can be found in many species and sometimes rather surprising ones, like that byword for harmonious cooperation, the honeybee. Honeybee queens emerge from special peanut-shell-shaped wax structures called queen cells in a beehive. If the colony is ready to swarm, then a newly emerged queen will inherit the hive once the old queen leaves with around a third of the honeybee workers. This is a very fine prize indeed but the worker bees like to spread their risk; perhaps a potential queen will die while metamorphosing from a larva to an adult? To reduce that risk, the workers rear more than one larva to be a queen, pumping the royal-destined grubs full of royal jelly after they hatch from fertilised eggs laid by the mother queen. The first queen to emerge typically finds the other yet-to-emerge queens and stings them to death through the wall of their queen cell. Her sting is adapted specifically for this purpose, being smooth like a dagger; she is an evolved killer of her own species. What is more, those victims are her sisters or half-sisters. Bumblebee workers, by the way, often kill their mother queen as the season starts to draw to a close, simultaneously committing matricide and regicide.

Conspecific killers abound in the animal kingdom. Lions are well known for killing their own kind when a new pride male takes over. Cubs are hungry mouths to feed and unrelated potential rivals when they become adults. Also, and a more important factor in the short term, females with cubs are not in oestrus, which limits the incoming male's reproductive opportunities. Infanticide provides a simple solution to all of these problems and a tasty canapé as a side benefit. Infanticide is known in a range of other mammals from gerbils to langur monkeys and in birds including wrens and jacanas (or lilytrotters, from their habit of walking over waterlily pads). Some particularly fascinating examples of infanticide can be found in insects. Giant water-bug males take care of eggs, guarding them and keeping them from drying out. Without males and their attention the eggs won't hatch and this behaviour makes males a valuable resource for females. If a female cannot find an egg-less male, then she stabs and kills the eggs of a brooding male. Facing a reproductively barren future, the male mates with the egg-stabbing female and takes her eggs to guard.

I could go on here, because there are very many examples to pick from, but the point is hopefully clear. Conspecific violence, even lethal violence, is not a uniquely human trait. What is unique is the way in which we've run with it. We've become the absolute masters of tearing lumps out of our own kind, an assertion that is backed up by a study published in 2016 examining 'sources of mortality' (causes of death) from 1,024 mammal species representing most of mammalian diversity.[2] A wide range of animals including whales, bats, pangolins and antelopes were thrown into the mix, along with a good mix of primates and more than 600 human populations ranging from the Palaeolithic era to the present day. Such an analysis is far from straightforward; it relied on collating data from a huge number of studies and, in the case of the human element, from a mix of sources that included archaeological evidence. Maria Gomez and colleagues defined the level of lethal violence as 'the probability of dying from intraspecific violence compared to all other causes'. They worked through the available evidence to calculate that probability as 'the percentage, with respect to all documented sources of mortality, of total deaths due to conspecifics'. This definition comes replete with a grim list that includes infanticide, cannibalism, inter-group aggression and any other type of intraspecific killings in non-human mammals, as well as war, homicide, infanticide, execution and other kinds of intentional conspecific killing in humans. They then looked at this data in relation to the phylogeny of mammals, the evolutionary family tree that shows the relationships of each group to the other. This phylogenetic approach allowed them to make solid inferences about the evolution of violence across the mammals as a whole. It's a neat and involved piece of work that made quite a splash when it was published in the world-leading science journal *Nature* in 2016.

Lethal conspecific violence was reported for a whacking 40 per cent of the mammal species in the study, and this is almost certain to be an underestimate because of a lack of data on a great many species. So, humans are very far from unusual in killing each other, and the evidence suggests that conspecific violence has evolved multiple times and has been an advantage in many different species and scenarios. They also found a firm connection between the level of violence and the degree of shared evolutionary history. This indicates that violence is more likely to evolve in certain groups, although they were also able to show that this has quite a degree of flexibility, leading to different patterns in some closely related groups. As ever, it's complex. Social species and territorial species were found to be substantially more violent overall, lending yet more weight to our speculative evolutionary scenario since we are social and in the most part highly territorial. Overall, some groups of mammals like bats, whales and lagomorphs (rabbits, hares and an adorable little group of fluffy creatures called pikas) had a low level of lethal violence, whereas other groups, most notably the primates, had a high level.

The analysis of lethal violence across mammal groups and knowledge of their evolutionary relatedness allowed Gomez and colleagues to predict the proportion of human deaths that would be caused by violence inflicted by fellow humans. Their prediction was 2 per cent and this value was robust to all kinds of changes in the statistical models used to generate it. Interestingly, 2 per cent was also pretty much the proportion inferred for the evolutionary ancestor of all primates (2.3 per cent) and of apes (1.8 per cent), indicating that lethal conspecific violence was a more or less consistent feature of our evolutionary past. As the researchers put it, 'These results suggest that lethal violence is deeply rooted in the primate lineage.' Across all mammals the level was around 0.3 per cent, making us around seven times more lethally violent to our kind than the average mammal. So far, our speculative evolutionary scenario is bearing up remarkably well.

There is of course a problem with values predicted from a statistical model and any inferences made from such values: they are simply predictions from a model. For humans though, we can test these predictions against actual values, because we have historical records and archaeological evidence. It is a tricky task to control for different factors including the absence of evidence from lethal attacks on preserved bones, the influence of battles, the likelihood of preservation of bodies that have died by different means and so on. However, controlling for as many of these factors as possible led the researchers to conclude that the level of lethal violence in human prehistory did not differ from the level predicted by their phylogenetic approach; it still comes out at around 2 per cent.

There was a large variation in the level of lethal violence between different periods and that allows us to infer certain things about the role of different factors in the expression of violence. Perhaps the most interesting of these inferences for our mismatch hypothesis concerns population density. Increased population density is a common driver of increased conspecific violence in other mammals but it seems that is not the case for us. Living in centres of high populations is, conclude the researchers, 'probably a consequence of successful pacifi-cation, rather than a cause of strife'. Modern life might actually be reducing our violent tendencies, which is a pattern that is seen in lethal violence across time both in the Gomez study and in other studies, although as we shall see this is disputed. Overall, the level of violence in prehistoric humans matches the phylogenetic expectation, and our ancestors were exactly as violent as our evolutionary heritage suggests.

A genetic basis for violence

The argument that we have evolved to be violent is holding up nicely, but if we are going to claim that 'evolution did it' then we need the smoking gun provided by genes; evolution, remember, is a genetic process. The convenient 'gene found for [insert trait here]' headline beloved of the media is rarely the case and there is no reason whatsoever for assuming that 'violence', a complex trait that must involve many different components, is influenced by a single gene. Also, and this cannot be stressed enough, violence isn't *caused* by genes. Having certain genes might lead to a tendency for violence to be expressed more in some people and in certain environments, but it is the interplay between genes and environment that is important in translating DNA to GBH.[3] And even then, violence can be controlled; you don't *have* to lash out. This is an important caveat to bear in mind because in fact it turns out there really are genes (note the plural) that are associated with violence (feel free to ring the correlation/causality alarm bells).

Some really clinching evidence for the genetic basis of violence in animals generally is the fact that we can breed animals to be more aggressive. Artificial selection for different traits, and the subsequent heritability of those traits in offspring, shows definitively that those traits have a genetic basis but rarely tells us much more than that. We know too that in humans, violence tends to run in families, as the stark title of a paper published in 2011 with an unusually large sample size conclusively shows: 'Violent crime runs in families: a total population study of 12.5 million individuals'.[4] Despite very convincingly backing up the title, this study still doesn't give us absolute clinching evidence for a genetic association, since families share physical and cultural environments. The fact that the study found differences in the likelihoods for different types of violent crimes (arson, for example, being especially likely to run in families) further underlines both the strong influence of environment and the fact that 'violence' is a complex multi-factorial phenomenon, complicating any hunt for genes.

Given the practical and philosophical need that we have to understand violence in the modern world, it is not surprising that considerable effort has been put into the topic and the scientific literature is well populated with both empirical papers searching for (and finding) genetic associations, and review and opinion pieces discussing their findings. Just recently, a study identified 40 genes related to aggressive behaviour and shared between mice and humans,[5] with genes for different neurotransmitter metabolising enzymes potentially indicating 'treatments' for aggression; the motivation being that if we understand the biochemical basis for violence, then perhaps we can mitigate it biochemically through drugs and other therapies.

We are starting to drill down into the genetic mechanisms of violence, and we will undoubtedly get an ever more refined and nuanced understanding of human violence as time goes by. For our purposes though it is enough to show that violence does have some genetic basis. Indeed a review of a number of studies looking at the genetics of violence concluded that 'approximately 50% of the variance in antisocial phenotypes is the result of genetic factors'.[1] In other words, genetics explain around half of the variation that we can measure in antisocial behaviour of which violence is a component.

Violence in humans is usually considered from a psychological and biochemical perspective. From this perspective, aggression and violence results from a state of mind developed through hormones and the nervous system, triggered by sensory input and mediated by conscious and subconscious processing within the brain. This is an entirely logical way to approach the topic. We are often interested in the differences between different people's tendencies to violence and we tend to associate them with emotional states like anger, temper and stress or with brain-related disorders like psychopathy. Studies of animals that engage in violence are far more likely to focus on the physical and behavioural attributes that facilitate violence and the ecological and social conditions that provoke it and that have contributed to its evolution. Ecologically, the Gomez study highlighted the role that sociality and territoriality play in the expression of lethal violence in mammals including humans, and the phylogenetic comparison of us with other animals turned out to be a useful one in disentangling some of the evolutionary features of violence. Other researchers have taken an even more 'zoological' approach, looking at how evolution may have shaped aspects of our bodies for violence.

Straight off the bat we can link physical attributes of humans to the ability to inflict harm on others, up to and including death. We are physically strong and fast and being bipedal leaves our arms free to swing around. Even an average person of average build with no particular training can generate enough power behind a fist to kill another person if the right spot is hit (remember our skulls are not especially heavily built in some places). We have vulnerabilities and weak spots: the groin, the stomach, the liver and the kidneys provide soft and painful targets, likewise the joints in our arms and legs are susceptible to debilitating strikes. A blow to the correct part of the neck, where the carotid artery sinuses are, can be an instant lights-out, while 'one punch' deaths are a not uncommon tragedy.[6] Our other limbs are no less lethal. Bony feet and powerful legs allow for the sort of graceful kicks you might see in the movies or the much more effective kind of brutal stomping that you might see in real life. Even though our skulls have vulnerabilities (the back of the neck, the thinner parts around the temple), the thicker portions at the front allow for our head to be used as a weapon for delivering devastating headbutts. On top of all this we can plan and use tools very effectively, allowing us to make use of all kinds of things as weapons, from sticks and stones to the Jason Bourne favourite, the tightly rolled-up magazine.

Make no mistake, we are a very handy physical unit when it comes to violence; but can any of the attributes we have for violence realistically be said to have evolved 'for' violence, rather than violence being just a useful by-product? The general opinion is no, but that hasn't stopped an interesting pocket of research developing that suggests that the ability to make a fist and punch someone without hurting yourself has been a driver in the evolution of the human hand. The history behind the idea is an amusing example of how scientific inspiration can hit. David Carrier, a comparative physiologist at the University of Utah, was arguing at a conference with other scientists about whether the sperm whale's oddly bulging head had evolved for the purpose of ramming other males in order to compete for females. Frank Fish, a biologist specialising in the biomechanics of aquatic mammals, thought not, illustrating his point by waving his fist and saying (according to Carrier) 'I can hit you in the face with this, but it did not evolve for that.'[7] Carrier was inspired; what if the hand *had* evolved for fighting? Or more accurately, the human hand, in the process of becoming more delicate and dexterous allowing for greater manipulation and tool-use, may also have evolved to ball into a fist that can punch but provide some protection against bone breakage.

Carrier's fist-balling hypothesis wasn't his first foray into the physicality of human violence and evolution. Some of his earlier work suggested that the human male face may have evolved relatively robust cheeks, jaw and brow to withstand punches. It is fair to say that this 'punch proof face' idea was not universally well received. Evolutionary biologist David Nickle was especially outspoken about it, saying in an article in the *Los Angeles Times*, 'What I find most objectionable is that this type of research does a disservice to the general public by supplying a very wrong understanding of human biology, and more generally, human evolution.'[8] The reason for these harsh words is that Carrier, Nickle thinks, is telling evolutionary Just So stories. Nickle goes on to add that 'I think Carrier and Morgan's [Carrier's co-author, United States physician Michael H. Morgan] argument is akin to arguing that human speech evolved so that humans could more effectively lie to each other.' This argument is an example of what has become known as a 'spandrel' evolutionary argument. Spandrels are the almost triangular spaces that are generated when two arches meet in a building and they are often used for the placing of decorative plaques or paintings. Evolutionary biologists Stephen Jay Gould and Richard Lewontin used spandrels as an example of a useful characteristic (a handy space to put some artwork) that is in fact merely a by-product of the 'evolution' of another characteristic (supportive arches). 'Spandrels' has become shorthand for the evolutionary loose thinking that a seemingly adaptive characteristic 'must' have evolved through adaptation rather than simply being a by-product; your fleshy outer ears are great resting places for spectacles but this is not the reason for their peculiar shape.

Carrier tested the fist-hypothesis using nine male arms that he obtained through post-mortem body donor programmes. Fishing line attached to tendons allowed Carrier and his team to control the wrist, fingers and thumbs of the arms. Strain gauges and accelerometers allowed the team to measure the force of impact, and the effects on the delicate bones of the hand, caused by swinging the arm against a padded dumbbell. It's a solid enough method but it must have looked rather macabre to anyone passing by.

Controlling the shape of the hand with the fishing lines in the style of a fleshy marionette, the impact and strain caused by a clenched fist, an unclenched fist (where the thumbs and fingers provide less protection) and an open hand revealed some interesting findings. A clenched fist, buttressed by the fingers and thumb in a way that is possible for humans thanks to our more dexterous hand, was able to deliver 55 per cent more force safely, without damaging the metacarpal bones, than an unclenched fist and can deliver twice the force of an open hand. Carrier concluded that the human hand and its ability to form a buttressed fist is an adaptation for fighting.

Once again Carrier found himself sharply criticised, with Frank Fish leading the charge both in the academic literature and in the press. Talking to the *Los Angeles Times*, Fish, who holds a black belt in taekwondo and so knows a thing or two about hitting people, pointed out that the human body has many parts that are useful for hitting people, like knees, elbows and feet, but no one is arguing that those evolved for fighting. In the scientific literature, Fish published a firm critique in the same journal that the research appeared, concluding that 'The *Homo sapiens* hand may have some fighting advantages, but there is no reason to believe it is anything more than adventitious happenstance.'[9] In other words, the 'fighting fist' is a spandrel. Others were more supportive of the idea. David Puts, a biological anthropologist at Pennsylvania State University, commented that 'I don't think that, by itself, it's convincing, but the authors are building an increasingly convincing case' while even Frank Fish conceded that 'I think he's moved the topic along.' It will be interesting to see where Carrier and others take this idea in the future. A possibly fruitful direction will be to examine differences between men and women, because as pointed out by Richard Wrangham, a biological anthropologist from Harvard with an interest in the evolution of violence, we would predict differences given that our underlying scenario is about male competition and violence supporting increased mating opportunity with females.

Controlling our tendencies
The evidence very much points towards an evolutionary explanation for human violence, or at least towards the tendency or possibility of lethal violence to be expressed as part of our behavioural repertoire. As unsettling as this might be, it is worth stepping back and taking a

broader perspective. We may all have the *capacity* for violence but most of us, most of the time, are not violent. We may think about violence, we may enjoy watching violence and we may wish violence upon others at moments of frustration or anger, but the reality is that very few of us actually visit violence of any kind on our fellow humans. Fewer of us still will ever kill another person in a deliberately violent act. Although the possibility exists, and we have the physical and mental abilities to carry it through, the vast majority of us are not killers. One explanation for this is that there are evolved, adaptive 'aggression inhibition systems' in the brain that prevent us from carrying out violence when the benefit is outweighed by the cost. When we are cut up in traffic for example we may feel rising anger and visualise beating the offender to a pulp but the risks of a hefty prison sentence, relationship breakdown and financial ruin, very real costs that we can calculate rapidly, might prevent us from acting out our revenge fantasies. In other words, we can exercise self-control and are not the helpless actors of our evolutionary script. Alternatively, if we or those close to us are at immediate and life-threatening risk then the cost–benefit seesaw might tip and even the mildest mannered individual might let rip and, in some circumstances, kill. Regardless of the precise mental mechanisms that control our violent tendencies, the evidence shows that most of us can. On the other hand, some people are very violent indeed and this mismatched pattern of violence in humans in the modern world warrants an explanation.

One explanation for extreme violence is simple variation. The genetics of violence, and the genetics that play a part in the mental structures that control our violent tendencies and thoughts, are complex and will differ between individuals. Whenever we have differences between people we inescapably have some people who end up at the extreme ends of the distribution. In this case, one end would be people that may be more violent than most and the other end are the peacemakers. It may be that by the mixing of genes to create offspring we sometimes bring together genetic combinations that result in very violent individuals, despite extreme violence not being adaptive in the modern world.

A focus on genetics, though important from an evolutionary viewpoint, is only part of the picture. We know that it is generally the case that the interaction between environment and genetics that leads to the expression of traits and violence is no exception. One model that accounts for this interaction, and which refines our initial evolutionary scenario, is the catalyst model, proposed by psychologist Christopher Ferguson. In this model humans have an evolved aggression inhibition system termed an impulse control device, likely in the frontal lobes of the brain. This is a component of 'self-control', the lack of which is a strong predictor of violent criminal behaviour. Self-control is, like violence, strongly influenced by genetics, which accounts for around 50 to 90 per cent of the variance that we can measure in self-control.[1] So far so genetic, but in the catalyst model it is the family environment (itself conflated with genetics of course) that interacts with our genetics to determine whether we end up with a violent personality. Environmental stresses and strains act as catalysts to provoke potential behaviours that are then filtered through the impulse control device. It is the relative influence of the drive for aggression (environmental experience interacting with genetic tendency) and the drive to inhibit aggression (via impulse control) that determines whether the environmental catalyst will result in a violent outcome. Some have interpreted this to mean that the frequency of violent behaviours will likely increase during times of higher environmental stress, and this does not seem like a particularly big leap intellectually. In layman's terms, if you put people under stress they are more likely to snap.

Given that our environment has changed dramatically over the course of our history and that we have very recently created a world with increasing levels of stress (Chapter 5) are we, as the headlines might have us believe, becoming more violent? It is a straightforward enough question; however, the answer is anything but. It is possible to find equally confident assertions that we are more violent than we were in the past,[10,11] that we are less violent, that violent crime is increasing, decreasing or staying the same, or that homicide rates have gone up or down.[12] Much depends on where you look, how you look, your definitions and your analytical approach. It would be nice to be on the side of American psychologist Steven Pinker, who posits in his 2011 book *The Better Angels of Our Nature: Why Violence Has Declined* that, as the title suggests, violence has decreased. His explanation focuses on the roles played by the emergence of nation-states with strong central governments, the development of stable and valuable trade networks and our ability to communicate, all of which increase our dependence on each other and reduce deaths due to violence. A central plank in that argument is provided by data that suggest fewer people in more 'modern' societies, relative to the society's total population, die in wars and conflicts than among the sort of small groups of hunter-gatherers and pastoralists that were typical for most of our history.

Pinker's positive viewpoint has not gone unchallenged. A team led by anthropologist Rahul Oka from the University of Notre Dame in Indiana, USA took a statistical approach to deaths in warfare over the course of human history and came to the conclusion that we are no more or less violent now than in the past. They reasoned that the number of people likely to be involved in warfare, a vital component of Pinker's argument, doesn't scale linearly with population. They assert, entirely correctly, that the proportion of the population engaged in military activities decreases as population increases. A band of 100 people might have 25 or even more people as a military force, for example, but a population of a million people would not have 250,000 ready and willing to serve in times of war. According to Oka and colleagues it is this simple scaling effect, and not increasing interdependence and the benefits of peace, that is the driver behind reduced proportional casualties of war as human societies become larger and more complex.

The role of the media in the modern environment

One of the environmental changes of recent years that is of great concern in our understanding and framing of violence in the modern world is the domination of violence on television, movies and gaming and its potential influence on children. We can add to this mix the recent rise of internet streaming sites like YouTube, where violent scenes, including real-life killings, executions and atrocities, are only a click away. It is important to be aware, if you are not already, that it is very easy indeed to find and view graphic scenes of beatings, shootings, beheadings and far worse online. This is not content hidden away on the 'dark web' but content that is freely accessible and searchable via Google and sometimes linked via mainstream news pages. The possibility of such 'entertainments' being a part of, and potentially a large part of, a child's environment as they grow up is real, recent and worrying. Such worries, although often stirred up by the very media that help to portray violence, are not to be trivialised. If we accept some form of the catalyst model, as seems sensible, then we accept the evidence that exposure to violence (including family and 'upbringing' violence) is a predictor of violent behaviour. If we also accept that our environment plays at least some part in determining the outcome of the tug-of-war between inherent violent tendencies and our inbuilt impulse control, then we must take seriously the idea that being exposed to violence at a young age could shape our response to violence. If we are creating an entertainment environment and online and virtual worlds where depictions of violence are interacting with, and amplifying, our evolutionary tendencies to violence, then we find ourselves in an extremely recent, concerning and potentially lethal mismatch.

Surprisingly, given that the links between media violence and real-world violence are often portrayed in the media as being 'controversial', there is relatively little controversy at all in the scientific literature. In 2011, the International Society for Research on Aggression, a society of scholars and scientists engaged in the scientific study of violence and aggression, appointed a special commission to prepare a report on media violence.[13] The links between media and violence are not likely to be straightforward and, like most of what we have discussed in this chapter, are a minefield of variation, conflated factors, correlations, causalities and more besides. As the report says:

> Of course, watching a violent movie does not normally lead people to assault another person when they leave the cinema. Nor is it true that avid players of highly violent video games often end up as violent criminals. No respectable researchers in this area would make such claims. Rather, the issue is whether watching violent movies and shows or interactively engaging in violent games in a virtual world increases the odds that people may engage in aggressive behavior in a variety of forms, both in the short term and in the long term.

With those caveats out of the way the findings of that report are clear and concerning. The commission found that 'After taking into consideration numerous characteristics of the child and the environment, including risk and protective factors, research clearly shows that media violence consumption increases the relative risk of aggression, defined as intentional harm to another person that could be verbal, relational, or physical.' In other words, there is a strong evidence base from an array of studies to support the hypothesis that 'monkey see, monkey do'. Mimicry, of course, is an excellent evolved ability found throughout the animal kingdom that allows individuals to pick up all kinds of life-giving skills; its co-option into developing violence in young people in a changed modern environment is yet another unfortunate evolutionary mismatch.

Cited in that report was the work of Albert Bandura who, along with others, carried out pioneering research in the early 1960s that examined children's behaviour following exposure to violent images. That description sounds rather less ethical than the experiments actually were, although getting the study past an ethics committee these days might prove to be a struggle. In what have become known as the Bobo Doll experiments, Bandura and others showed some children film of an adult playing with an inflatable doll during which the adult hits the doll with a mallet, kicks the doll, sits on it and generally gives poor Bobo a pretty hard time. After the viewing, children were taken to a playroom that contained lots of toys and, crucially, a Bobo doll. The children who had seen the film of Bobo getting a good pasting didn't just recreate the violence they saw; they came up with new ways to beat up Bobo, and they played more violently with other toys. Seeing aggressive behaviour then isn't just a case of 'monkey see, monkey do'; it is 'monkey see, monkey inspired'. Witnessed violence isn't just mimicked, it is enhanced and refined, seeming to unlock and release the violent potential within. Fast-forward 50 years to a time of immersive ultra-violent video games, 3D virtual reality and streaming violence 24/7, and suddenly poor Bobo getting smacked over the head with a mallet seems like the mark of a gentler, more innocent age.

In the short-term, exposure to media violence causes changes in our brains. Things that are experienced at the same time start to become linked through neural connections, like for example a warm loaf of bread and its distinctive smell. As the connection strengthens between these 'nodes' through multiple exposure, the activation of one node can lead to the partial activation of the other in a process called spreading activation. Connections between strongly linked things can develop rapidly and even in young children there are neural pathways between indicators of aggression (guns, shouting) and violence (fighting, hitting). The consequence of these connections, which are an evolved part of the way our brains function, is that if a person is exposed to a violent scene, the resulting activation spreads out to connected nodes and activates, or primes, them slightly. When nodes associated with behavioural tendencies are primed, it makes the behaviour more likely to be manifested. If you doubt the power of priming, then consider this: studies show that if someone is being insulted, then the mere sight of a gun can cause the insulted person to lose self-control and retaliate aggressively. The sight of a gun has a priming effect, the insults activate the aggressive behaviour node further and the thin veneer of docility provided by our impulse control disappears rapidly. Exposure to media violence has the power to prime a very wide variety of nodes.

In the long-term, exposure to violence may have more subtle and disturbing effects. Our lives are complex and to make sense of that complexity we develop what psychologists call schema or scripts. In certain scenarios, like visiting relatives, a complex diversity of emotions, concepts and sensations are activated together in the brain. Perhaps you are visiting your grandmother's house and the journey there, the sight of the house, its distinctive smell, the feelings you get on seeing your grandmother and so on are activated and they themselves link to certain memories and to a specific mental 'script' that allows us to behave appropriately in that situation. The ability to link together appropriate emotions, sensations and memory, and to act accordingly, is a valuable skill for a social animal. What is important to bear in mind is that once these 'scripts', these integrated knowledge structures, are activated they are key determinants of how we behave and they can influence what we do outside of what we are aware of consciously.

The schemas and scripts that influence, modify and to a great extent control our behaviour are affected by our experiences. They are moulded by the world around us as we grow and develop. Parents, siblings and other family members usually form our primary nurturing environment but as we develop other influencers allow us to modify our scripts and learn new ones, such as the 'being at school' script. For the child growing up in the modern world the influencers have developed beyond parents, peers and schools to include the media. If being with parents and peers is sufficient for children to learn appropriate behavioural scripts for situations, then being 'with' media is no different. There is now no real doubt that children learn from what they see and experience on screens, and if what they see is violence then their developing neural networks will reflect that.

Media violence can write new scripts for us but it has other effects. Violent scenes can act as triggers for violent thoughts and feelings already stored in memory and, if activated sufficiently often enough, can influence behaviour. It is also suggested that repeated viewing of violence and activation of those nodes in the brain can make people more prone to interpreting ambiguous actions as deliberate provocation: the 'did you spill my pint?' scenario. Add to this the portrayals of violence as 'fun', the fact that violence is often rewarded or respected and the framing of violence as heroic, and then we have a potent mix indeed for influencing our behaviour. Video gaming merely exacerbates this effect, with the repetition of violent acts resulting in changes to the brain that make it more likely for people to behave aggressively in the real world; when gamers play violent games, aggressive attitudes and behaviours are more likely (but not inevitable) to cross over from pixel to playground.

Another concern over the portrayal of violence in modern media, and in this we can include the news and other 'worthy' media as well as more recreational viewing, is its potential for 'desensitising' us. We watch the same thing over and over again, then we are no longer as emotionally engaged with it and, through this exposure to repeated stimuli, we become desensitised, no longer responding in what we would previously have regarded as an appropriate way. Whether it is mass shootings, famine, war or disease this desensitisation is well known and relatively well characterised. Desensitisation is another example of evolutionary mismatch with the modern environment. It is actually commonplace and important in our lives; the reduction in cognitive, emotional, physiological and potentially behavioural responses to certain stimuli protects us and allows us to carry out tasks or be in situations that would otherwise be difficult or impossible. Such situations don't have to be dramatic or violent; prior to becoming a parent, dealing with a bed full of diarrhoea and vomit would have elicited a highly emotional and possibly physiological reaction from me, but experience has led to a degree of desensitisation, allowing me to deal with such situations relatively calmly and effectively. Desensitisation also explains 'donor fatigue' and the diminished power of emotionally charged 'poverty porn' images to elicit money for charities.

The modern world is characterised to a great extent by the pervasiveness of electronic media. Violence, including real-world lethal violence, sadistic violence, rewarded violence, 'heroic' violence and violence that can be acted out by the viewer as a willing participant is ubiquitous and readily accessible. The connection between media portrayals of violence and subsequent aggressive behaviour in children is well established on foundations laid down in the early 1960s. With ever more sophisticated gaming and highly accessible media on devices that never leave our sides and that attract children like moths to a flame, this new environment is a very clear potential mismatch with our evolutionary history if we want to live in a less violent world. It is something that should concern us all, as the children that developed their mental scripts and aggression inhibition systems in the last decade or so become adults. In case we feel like playing the ever-popular 'blaming the next generation' game, there is also increasing evidence that adults are affected by watching violence, making them more aggressive, although only if they tend to have more aggressive personalities to begin with.[15] This should make us doubly worried given what we know about the influence of violence on developing mental scripts in children. We should always remember though that violence on the media isn't the cause of violence; we were the most lethally violent mammal long before TV was invented.

Reducing violence

Discussing violence from an evolutionary perspective can attract criticism, in part because it is all too easy to think that an evolutionary argument 'for' a trait in some ways justifies that trait. This is an understandable reaction, but developing an evidenced idea that violence is in some way a part of our evolutionary heritage in no way excuses violence. We can develop, and indeed have developed, an equally powerful evolutionary argument for impulse control and the reduction of violence; it is this side of the complex balancing act that wins through most of the time in most of us. Understanding the evolutionary origins and significance of violence, and understanding the environmental triggers that provoke violence, actually gives us powerful insights and opportunities to understand and deal with violence. As Christopher Ferguson argues in a paper with Kevin Beaver entitled 'Natural born killers: the genetic origins of extreme violence':

> From behavioral genetics and evolutionary models of violence, we may more fully understand which individuals

are at greatest risk for extreme violence. We can then begin to examine the interaction not only between genes and environmental catalysts for violence, but also the interaction between genes and treatments and prevention efforts for violence.

The hope is that an evolutionary understanding might lead to more effective interventions to reduce violence, and that can only be a good thing.

CHAPTER EIGHT

Hopeless Addicts

Humans love drugs. Give us something that causes some alteration in our state that we find pleasing and we can't get enough of it. We smoke tobacco, drink alcohol, roll spliffs and joints of cannabis, snort cocaine, drop acid, shoot heroin, mix speedballs, pop pills, make meth, chew coca leaves, trip out on mushrooms, brew coffee and countless other activities that get us high, wasted and otherwise 'state altered'. I'll be honest here, stranded on a desert island I'd be getting the first batch of coconut wine on the go before I even thought about building a signal fire.

Human history is a litany of drug-fuelled escapades. Genghis Khan's son Ogedai died from an excessive consumption of wine. The Aztecs were keen users of *Psilocybe mexicana*, a psychedelic mushroom. Both the Aztecs and the Maya consumed the seeds of the morning glory plant that contain ergine, a compound related to the powerful hallucinogen LSD. The blue lotus flower, a beautiful water lily, was sacred to the Egyptians as a symbol of the sun but it also contains a psychoactive substance called aporphine that was also known to the Maya across the Atlantic. In more recent times, Britain went to war with China twice over the opium trade in the 1800s. Chewing on plants or smoking bits of them to get wasted is one thing, but once the technology of alcohol brewing got under way we could get our buzz on whenever we liked. Exactly when we first realised that we could make booze is much debated, but the discovery of a 13,000-year-old residue of beer in a cave near Haifa in Israel strongly suggests we've been at it for a fair time. The burgeoning gin and craft ale markets are strong signs that we're not stopping in our quest to get state-altered.

The ubiquity of drug use across human culture was noted by the German scientist Baron Ernst von Bibra. Von Bibra was quite something; he was one of those nineteenth-century characters you come across every so often that make you wonder how you've never heard of them before. Born in 1806 in present-day Bavaria, this remarkable polymath published on a dazzling range of topics, most of which he helped in some way to define, including industrial chemistry, archaeological metallurgy, chemical archaeology, zoology and botany. He was fortunate, or particularly talented, to survive 49 duels as a young man and, perhaps not coincidentally, went on to develop a special interest in narcotics. In a book examining the cultivation, preparation and consumption of various plant-derived intoxicants from around the world, *Die narkotischen Genussmittel und der Mensch* (translated as *Plant Intoxicants: A Classic Text on the Use of Mind-Altering Plants*), he notes that: '… there exist no people on the Earth that fails to consume one or another of these dainties, which I have subsumed under the name "pleasure drugs". There must therefore exist a deeper motive, and the notion of fashion or the passion for imitation cannot be applied here.'

The search for that deeper motive, and how it conflicts with the modern world in ways that are both spectacularly mismatched and incredibly harmful, is what this chapter is all about.

The burden of drugs

The costs of drug-taking are great. If we reduce it to the financial burden, and we accept that putting a cash value on some of the negative effects of drugs is problematic but not impossible, then a recent estimate put the cost just in the USA at a staggering $820 billion (£650 billion) a year.[1] The cost of alcohol abuse in the UK alone is estimated to be £55.1 billion.[2] We can subdivide this figure to give more of an idea of the breadth and depth of the burden caused by our love for booze: £22.6 billion is the cost to individuals and households through crime, violence, private health care and lost income due to unemployment; a similar amount goes on the human cost of pain and grief; £3.2 billion on NHS healthcare; £5 billion on social care, fire services and criminal justice services; and £7.3 billion on lost productivity and absenteeism. Smoking tobacco accounts for 2 per cent of the world's entire gross domestic product and kills more than 2 million of us every year.[3] Moving to 'harder' drugs, more than 70,000 Americans died from drug overdoses in 2017. That figure is up nearly 10 per cent from the year before and the dramatic rise is being driven by a 13-fold increase in the use and abuse of synthetic opioids over the decade from 2007.[4] Opioid painkillers available both on prescription and illegally include oxycodone ('hillbilly heroin'), hydrocodone (Vicodin, favoured by Hugh Laurie's TV character House MD) and fentanyl, from which musician Prince died of an accidental overdose in 2016. The abuse of opioids in the USA is described variously as 'a crisis' and 'an epidemic' and has sparked legislative moves to prevent and treat addiction. Despite the declaration of a State of Emergency by the United States Department of Health and Human Services in 2017 and the senate-passed Opioid Crisis Response Act of 2018, the crisis continues. The abuse of drugs is, without any doubt at all, a massive burden on societies around the world.

Against this terrible backdrop of human suffering, financial cost and death it is not surprising that we have tended to become conditioned, at least at a societal level, into thinking that 'drugs' are 'bad'. Such thinking isn't the consequence of some recent awakening. The Gin Act of 1751 was enacted to reduce spirit consumption, which was seen as a major driver of crime in London. The Temperance movements of the nineteenth century fought a ceaseless battle against the bottle. In recent times, we have the 'war on drugs' and countless public health warnings, advertising campaigns and police crackdowns. Our feckless use of drugs is clearly endangering our personal health and well-being and having serious knock-on effects to wider society, but can we once again seek an explanation in a mismatch between our evolutionary heritage and the modern world we have created? The answer is a solid yes, but to understand why we first have to take a far less judgmental view of drug taking.

Drugs are bad, but oh so good...
'Drugs are bad' is a society-level way of thinking. At a society level, such a statement is clearly correct and nothing I am about to write will suggest otherwise.

Nonetheless, that mind-set conflates the effects that drug consumption has on society with the effects that the drugs themselves have on individual drug users, especially in the short term. If we are to understand why we take drugs from an evolutionary perspective and why the modern world is such a mismatch for the evolved beasts that we are, we need to take a more honest and mature perspective; we need to accept that drugs make people feel good. Really good.

The taking of state-altering chemicals, from Heineken to heroin, from weed to wine, or from cocaine to caffeine, changes the way that people feel. Depending on the substance taken the user may experience feelings of euphoria, elation, invincibility, hyper-alertness, strong emotional bonds with other people (especially fellow users) and the removal of pain, anxiety and doubt. We may become more alert, our senses more attuned, or we might slip into a state of extreme relaxation or a richly tapestried dream-filled sleep. These strong and usually desirable changes in state mean that taking drugs, at least initially, is a very much a good thing for the user in terms of their enjoyment of life. Some drugs might cause users to experience hallucinations and other strong psychological effects. Much is written about the terrors of hallucinations and 'bad trips', but many more people take good trips that enhance rather than reduce their life experience.

It is all too easy for the word 'drugs' to become shorthand for 'illegal' and 'harmful' substances like cocaine, heroin and crystal meth, but legally and widely used alcohol has strong and pleasurable effects that are familiar to most people. The line between what is socially acceptable and legal in terms of state-altering substances and what is socially unacceptable and illegal is increasingly blurred by the debate over the legal status of cannabis and so-called legal highs. The drawing of this line is a matter for politicians (ideally, but not always, informed by science-based evidence), but regardless of where it is drawn we will continue to take drugs because taking drugs makes us feel good. They even make us feel good in the face of sure and certain knowledge of the harm they can do both to some individuals and to society overall. Bans, laws, draconian judicial measures, overdoses and human ruin have never, and will never, stop us from taking drugs. Because, and it's worth saying this again, they make us feel good.

The 'reward pathway'

From an evolutionary perspective we can understand why we continue to take drugs, despite the knowledge that they could do us harm in the long run, by considering what pathways in the brain are stimulated by drugs and examining what other stimuli evoke similar responses. Fortunately, despite many drugs being very different chemically, there is a tendency for many of them to be rather similar in terms of their effects on the brain. Specifically, many drugs have an effect on a pathway called the mesolimbic pathway or, as it is more descriptively known, the reward pathway. This pathway, when stimu-lated, releases dopamine (a neurotransmitter that we've met before in Chapters 4 and 5) that has multiple effects. First, it regulates what is termed 'incentive salience', or the desire and motivation for rewarding stimuli, more of which shortly. Second, it facilitates reinforcement learning, allowing the brain to form connections between stimuli and reward. It may also be involved in the subjective experience of pleasure.

The rewarding stimuli that are usually involved with the mesolimbic pathway fall into two categories.

- Intrinsic, primary stimuli are those things that are good for survival and the production of offspring; sugary food and sex fall neatly into this category. The mesolimbic pathway has been selected for and evolved as a mechanism to incentivise us, and to train us, to seek out things that keep us alive and reproducing.
- Extrinsic stimuli are things we have learnt through the secondary actions of the mesolimbic system to associate with pleasure, like watching a favourite team win or buying a new pair of shoes.

Ultimately the mechanisms involved in drug taking are biochemical, with substances acting at a molecular level in our brains, but for the user the effect is 'felt' at an emotional and physical level. Taking drugs stimulates us to feel, often in a highly magnified way, the euphoria and excitation that we have evolved to feel in response to stimuli that were vital for our survival. The 'high' we might get from a successful hunting expedition, from a fruitful foraging trip or from sex are evolved responses, producing a pleasurable sensation that we recreate by repetition. When that high is replaced, and greatly accentuated, by the high we get from drugs our brains respond in much the same way and we seek out more of the same.

This explanation for why drugs are so effective is known as the 'hijack hypothesis'. The idea behind the hypothesis is that drugs 'hijack' our mesolimbic reward system, generating a signal that the drug taking is a behaviour conferring a fitness advantage. Anything that feels this good can't be bad, right? The hijacking of pre-existing, evolved brain machinery is a powerful and intuitive explanation for drug taking that frames our modern world behaviour in a firmly evolutionary context. It also explains drug-seeking behaviour in non-humans.

Many species have been known to develop more than a taste for alcohol, from close relatives like chimpanzees to far more distant relatives like horses. When it comes to harder drugs, rats have proved to be especially useful models for studying their effects. We already know a great deal about rats, they are relatively easy to work with and, like us, they are absolute drug-hoovers if given the opportunity. Rats can be connected to a drug dispenser that will provide a measured dose of whatever drug you wish to test in response to the rat pushing a bar in the cage that dispenses the drug. The rat pushes the bar and gets a hit, perhaps of cocaine. The drug is dispensed down a tube that can be connected directly to the nucleus accumbens in the brain. The nucleus accumbens is a region at the base of the forebrain that is an important part of the mesolimbic, or reward, pathway. When the rat presses the bar, the cocaine is mainlined directly into the brain pathway where it causes the release of dopamine. As we have seen, not only does this dopamine release stimulate the reward pathway, it also reinforces the behaviour of taking the drug, teaching the rat's brain to repeat the bar pushing.[5] And repeat it they do. Rats will enthusiastically learn to self-administer cocaine in this experimental set-up and the brain pathways being activated are identical to those in our own brains. The same result is found if the rats are connected up to a dosing machine loaded with heroin or morphine, again dispensed straight into the mesolimbic pathway. The message from these sorts of experiments is very clear; from an evolutionary perspective, rewarding some behaviours is good. As far as our brains are concerned, drugs provide very effective rewards.

From use to abuse and addiction

The continuation of drug-taking behaviour, whether in humans or rats, can result in a state of addiction. Addiction can become a complex topic both biologically and sociologically, with definitions of what is or isn't 'addictive' combining with confusion between addiction and dependence to create important but tricky problems. My natural tendency would be to swerve the topic altogether and move on to the special challenges presented by the modern world, especially given that the hijack hypothesis seems to so neatly explain our drug-taking tendencies. However, addiction is one of the major costs of drug taking, whichever way we choose to quantify it. What is more, although I have focused on drug taking there are many other aspects of human behaviour that involve, at least in everyday language, 'addiction'. We are increasingly seeing a narrative develop around mobile technology use, for example, that frames our incessant and seemingly out-of-control use of it, and of the social media supported by it, as addiction. Gambling can also lead to addiction and this has, as we shall see, plausible roots in our evolutionary past. Sex, sugar, even shopping, can also fall into being 'addictions' and, when they reach that level, can become highly problematic for people unable to control and change their behaviour. The hijack hypothesis neatly explains why we take drugs, and the learning of extrinsic stimuli in relation to our reward pathways might explain why we pursue other potentially harmful behaviours, but in the context of the modern world this is only the start. The harm of drug taking (excluding early overdose) generally comes once the behaviour is well established and an individual graduates from 'user' to 'addict'. This step has to be understood if we are to understand fully just how badly the modern world we have created interacts with our evolutionary past, not just for drugs but for other behaviours that are of increasing concern.

Having said that addiction is a complex problem, the biological connection between drug use and drug addiction is refreshingly simple. Addiction is generally defined as a disorder of the brain's reward system, characterised as a compulsive engagement with rewarding stimuli despite adverse consequences. In other words, the taking of drugs that stimulate the reward pathway stimulates the release of dopamine that helps the brain to learn to 'do more' of the behaviour being rewarded; a destructive, positively reinforcing feedback loop. A major characteristic of addiction is the breakdown of self-control over intake (we have met self-control before in Chapter 7). Cravings develop through interactions in several brain regions[6] and subsequent interactions with the reward pathway. Those cravings push the addict towards behaviours that will bring forth more of the drug but also of course towards the behaviours that society ends up shouldering as the costs of drug addiction: robbery, burglary, violent crime and so on (with clear links to issues raised in Chapter 7).

The repeated use of drugs trains our reward system and can lead to addiction. Some drugs, though, can have other more drastic neurological effects. With repeated use of heroin, for example, users can become dependent. Dependence develops because repeated use means that neurons within the thalamus and the brainstem, separate and different from the reward pathway, adapt to functioning in the presence of that drug. These neurons become unable to function normally if the drug is absent. If the user stops taking the drug in question then they will experience the unpleasant suite of symptoms that create 'withdrawal'. These are physical symptoms that vary with the drug. The classic heroin 'cold turkey' symptoms include extreme anxiety, sweating, shaking, vomiting and diarrhoea and are brought about by the sudden and complete withdrawal of the drug rather than a carefully dose-controlled reduction of intake. Withdrawal can be medically serious, and the sudden withdrawal of some substances upon which a person has become dependent can be life-threatening (as is the case for example with benzodiazepines and barbiturates). Withdrawal symptoms typically follow a pattern of feeling dreadful, and then feeling much worse, until the withdrawal is complete. The only way to alleviate those symptoms is to take more of the drug that has resulted in dependence, which of course combines with addiction (via the reward pathway) to create a very strong drive indeed to get hold of that drug. It is that strong drive that is at the root of much of the crime-related cost that drug taking imposes on society.

Addiction and neurological dependence are physiological conditions predicated on well-characterised interactions between certain substances and our brains. Complementing these physical conditions, psychological dependence is also a strongly motivating force in drug-taking behaviour. Symptoms of psychological dependence include panic attacks, dysphoria (a profound sense of unease or dissatisfaction), a reduced motivation or ability to experience pleasure (known as anhedonia) and our old friend stress (Chapter 5). Psychological addiction is also based on biochemical processes; ultimately everything we think or feel is chemical. Dependence is likely caused by changes in neurotransmitter activity or changes in the way that receptors in the brain work. Once again, dopamine seems set to play an important role.

It is all too easy to think of dependence and withdrawal in terms of 'hard' drugs, illegal substances like heroin causing the agonising writhing of threadbare junkies in some filthy badly-lit drugs den. However, dependence and withdrawal are not always so 'exotic'. In 2007 an article in the medical journal *The Lancet* compared the harm of drugs by devising a scale that scored the physical harm, the likelihood of dependence and the social harm of 20 drugs using nine categories scored from 0 to 3 by an expert and experienced panel.[8] Top of the shop was heroin, then cocaine, barbiturates and street methadone, an opioid used under prescription to manage the symptoms of opioid withdrawal but which is itself an addictive opioid with a street value and abuse potential. However, coming in at number five is the drug of choice for a great many of us in all walks of life: alcohol. Alcohol dependence and withdrawal are so serious that, like benzodiazepines and barbiturates, sudden withdrawal can be life-threatening. At number nine, just one place below amphetamine, is another everyday drug of choice, tobacco, that scores higher than cocaine for physical dependence (1.8 out of 3 compared to 1.3 for cocaine) and just 0.2 behind cocaine in terms of psychological dependence (a whopping 2.6 out of 3 for tobacco and a slightly higher 2.8 for cocaine).

Politics as 'environment'

The paper's lead author was Professor David Nutt, an English neuropsychopharmacologist, a job title that would be difficult to say after too many glasses of wine. Nutt specialises in the study of drugs and their effects, including addiction. His work is worth a short digression here because it tells us something about the modern environment in which we, as 'pre-adapted drug-users', operate. That environment includes, as we've seen in Chapter 7 and will see in Chapter 9, the media but it also includes the political climate in which we live.

If the name David Nutt is familiar to you then it's likely because he is no stranger to controversy. In 2008 he was appointed as the chairman of the Advisory Council on the Misuse of Drugs (ACMD), an advisory public body in the UK that was formed in 1971 following the Misuse of Drugs Act passed that same year. As chairman of the ACMD, Nutt found himself at loggerheads with a government more concerned about making political capital than considering empirical evidence. In 2009 Nutt published an editorial in the *Journal of Psychopharmacology*

entitled, 'Equasy – An overlooked addiction with implications for the current debate on drug harms'. If you
haven't heard of equasy, don't worry, it's not that you need
to become more attuned to the latest shifts in societal
drug habits. Nutt was being deliberately provocative in
the editorial by comparing the risks of horse riding
(horses being equids) with ecstasy (3,4-methylenedioxy-
methamphetamine or MDMA), hence equasy or Equine
Addiction Syndrome. In the editorial, which is well worth
a read, Nutt outlines the considerable risks and harms
associated with horse riding and concludes that there is a
'serious adverse event' every 350 exposures.[9] The harm
caused by horse riding can include serious paralysis and
even death (10 people per year). Ecstasy on the other
hand has one adverse event every 10,000 exposures. Using
the scale used in the 2007 paper, Nutt concludes that
horse riding is far more harmful than ecstasy and yet
society feels no need to control this activity.

Nutt wasn't suggesting that horse riding should be
banned. Instead he proposed that we take a mature and
rational approach to the actual risk of harm from drugs
rather than the perceived risk of harm, the portrayal of
which may be politically motivated (the 'war' on drugs
being a vote winner on many levels) and exaggerated by
the media. The highly biased nature of media coverage
of drugs deaths is highlighted by a study, cited by Nutt
and published in 2001, which showed that only one in
every 250 deaths by paracetamol in the preceding decade
was covered by the media but that every single ecstasy-
associated death got the full media treatment.

Nutt's editorial calling for a rational, evidence-based
approach did not play out well in a political environment

already primed for yet more legislative control of drugs. In returning cannabis to the status of a Class B drug, the then Home Secretary Jacqui Smith had, the year before, gone against the recommendations of the ACMD who had voted (by a majority of 20 to 3) to retain its Class C status.[10] By 2009 Smith and others were pushing to keep ecstasy as a Class A drug despite evidence suggesting that the harm it caused did not warrant this status. In February 2009, following Nutt's editorial, Smith called for him to apologise for his (factually accurate) comments that more people died from horse riding every year than ecstasy, a position that was described as 'bullying' in the *British Medical Journal*. In October 2009, a pamphlet was published containing a lecture that Nutt had given in the July of that year in which he had reiterated his view that drugs should be classified using an evidence-based approach underlined by the actual harm they cause. This was the final straw for a government invested in a public image of being hard on drugs and the new Home Secretary Alan Johnson dismissed Nutt from his role. Nutt may have had a challenging 2009 but things did not go well for Smith that year either. Embroiled in the MP expenses scandal over her true place of residence, in February it emerged that she had claimed expenses for a telecoms bill that included pornographic films watched by her husband. Never let it be said that the environment we have created in the modern world is simple...

Potency and availability

The hijack hypothesis explains, at a neurological and evolutionary level, our penchant for drugs but on its own it does not explain any potential 'mismatch' between evolution and environment. To address this, we need to look at the other side of the equation: our modern environment. As we've already seen, when it comes to drugs we can subdivide 'environment' in a variety of complex ways that include wider social contexts and the political institutions under which we live. Rather than focusing on the role of politics and society in controlling drug, taking, and getting enmeshed in arguments well away from biology and evolution, I would like instead to consider the environment as defined by the drugs themselves, specifically the types, strengths and availability of drugs in the modern world.

We have already seen that humans have a longstanding, rich, varied and well-documented relationship with drugs, but what becomes clear from the study of the history of drug taking is that our ancestors had shocking limitations of opportunity compared to the modern human. The drugs most associated with addiction, harm and problems in the modern world simply weren't available to our ancestors. Consider cocaine. It is widely known that indigenous peoples of the Andes mountains have long chewed on or brewed tea from the leaves of plants in the genus *Erythroxylum*, notably *Erythroxylum coca*, and continue to do so in many areas. Coca leaves contain cocaine and chewing on them gives a number of benefits to users including mild stimulation, suppression of hunger and thirst and some relief from altitude sickness. However, don't go thinking that your average Andean resident is talking fifteen to the dozen about anything and everything, embarrassing themselves at some middle-class dinner party. The amount of cocaine in coca leaves is very small, coming out somewhere around 0.5 per cent by weight depending on species, variety and location.[11] Cocaine as we know it in the modern world is a very different drug altogether.

To get from cocaine in coca leaves to cocaine hydrochloride, the substance we identify as pharmaceutical cocaine, is an involved and labour-intensive process. To start with you need an awful lot of coca leaves. To make a kilogram of cocaine requires half a tonne or more of leaves, which is rather more than the odd bag you might find in an Andean market. Once you have procured your raw ingredient you need to extract the cocaine from it using either solvent extraction (commonly with petrol or kerosene) or acid extraction using sulphuric acid. Both methods require the addition of a number of other chemicals to allow the coca paste to come out of solution, or precipitate. The resulting paste, known as cocaine base, is further worked chemically with hydrochloric acid and physically using hydraulic presses and filters to create cocaine hydrochloride, which can be made into a powder by being heated.[12] The resulting powder is typically taken by insufflation, better known as snorting, whereby the powder is inhaled into a nostril via a suitable tube or rolled-up bank note.

Remember, coca leaves contain around 0.5mg of cocaine per 100mg of leaves, and to make a kilogram of cocaine can require anything up to 500kg of coca leaves. Typical doses of cocaine hydrochloride as snorted by an average recreational user are around 50–100mg per dose. Assuming a purity of 50 per cent (street cocaine is increasing in purity and 60–80 per cent is common[13]) a typical user might intake, in one dose, around 25 to 50mg of cocaine. Now, not all that cocaine is biologically available and there are other factors to consider but already we can see that, in broad-brush terms, you'd have to consume an awful lot of leaves to get even close. Not allowing for any differences in availability or other factors, we are talking about consuming 10g of leaves in a few seconds to get a similar hit. For reference, a decent handful of privet leaves from my back garden, not dissimilar in terms of size and shape to coca leaves, weighs around 10g and eating these quickly would be a struggle. In fact, coca leaves tend to be chewed slowly throughout the day and while some users can end up taking in an impressive 200mg over 24 hours, the concentration of intake is very low and this high level of use is unusual.[14] The concentrated, intense and pleasurable hit that cocaine hydrochloride provides is only possible through the conversion of naturally occurring leaves into highly processed powder.

Likewise, the other well-known and highly addictive drug, heroin, is quite far removed from the natural product from which it is manufactured. Opioids derive from opium, the dried latex sap that is exuded from cuts made in immature flowering heads of the opium poppy. The latex contains morphine along with other substances including codeine and thebaine, the substance from which oxycodone and other opioids involved in the US opioid crisis can be manufactured. As with cocaine in coca leaves, we have a long history of opium use dating back more than 6,000 years, with evidence of opium being found from that period at the Cueva de los Murciélagos (the 'bat cave') in Andalucía, Spain. Morphine is the most abundant, powerful and important of the substances in opium and it makes up around 10 per cent by weight of the latex. Opium then is a natural product with a long history of human use and with a potent effect. The potency of natural opium pales though when compared first with morphine, extracted from opium in the early 1800s, and then with diamorphine, or heroin, first synthesised from morphine in 1874. Morphine, concentrated from the 10 per cent found in opium, was quickly found to have a much more potent effect and to be more addictive than opium, while heroin (a trade name devised by the chemical company Bayer) is considered around twice as potent again as morphine.

Cocaine and heroin demonstrate an important feature of the drugs environment of the modern world: drugs are much more potent these days than they were. We have refined and developed technological methods to process and concentrate naturally occurring substances to produce powerful drugs, but the increase in drug potency in the modern world can also be seen with a drug that exists in exactly the same form now as was enjoyed by our ancestors. Cannabis, whether in the form of buds and leaves of the *Cannabis* plant or as the resinous extract that makes hashish, is typically heated or burnt and the smoke inhaled. You can also eat it or brew a tea from it. Whether you chuck it on a fire or develop extravagant bongs and vaporisers to get the smoke into you, the basic scenario is the same as it was thousands of years ago. Cannabis has not, so far, been industrially transformed into a different, widely available product as was the case with coca and opium. What has happened with cannabis, though, is a form of evolution via artificial selection. We have selectively bred varieties of cannabis plant that produce an end-product with dramatically more tetrahydrocannabinol (THC) than is found in the naturally occurring plants available to our ancestors. A large-scale systematic review and meta-analysis of THC content in herbal cannabis (the buds and leaves of the plant) showed an overall increase from around 1–2 per cent in 1970 to 5–9 per cent in 2009, with an overall average increase of 5 per cent. The trend over the next decade was for this increase to accelerate, with cannabis getting stronger at a greater rate in recent years. Some strains of cannabis regularly test with THC concentrations well over 20 per cent[15] and the legalisation of cannabis in some parts of the USA has led to the development of cannabis products with very high THC levels.[16] The recent cannabis environment is beginning to change in a similar way to cocaine and opium, with technology creating far more potent products. Together with breeding more THC-rich strains of cannabis, it is a development that is concerning many who fear that the increasing use of high-potency cannabis can lead to psychosis, a connection that is far from straightforward but increasingly supported by evidence.[17]

As well as increasing potency, the drug environment of our recent past has also seen a step change in the range of drugs available to us. We have done more than just improving on drugs that we were already taking; we've made completely new ones. A long list of synthesised abused drugs includes amphetamine, methamphetamine ('crystal meth'), ecstasy, barbiturates, LSD, phencyclidine (PCP, or 'angel dust') and ketamine. Meanwhile, globalisation has greatly increased the ease with which drugs can be shipped around the world and the internet provides all kinds of potential for buying and selling. The modern environment is one of great temptation and a wealth of new and potent opportunities to highjack a brain that evolved to enjoy, and reward, the pleasures of sex and ripe berries.

Drunken monkeys

It is a relatively uncontroversial thing to say that a great many people have sex while drunk or at least while under the effects of alcohol. It follows that unless there is some mechanism whereby drunk people can't conceive, and the problem pages of many magazines strongly suggest this is not the case, many of us were likely conceived through an act of union inspired and facilitated by alcohol. Alcohol lowers inhibitions and can make us, in the short term at least, more socially outgoing. Recent studies also suggest that cannabis use might have positive effects on sexual frequency[18] while ecstasy has been shown to increase sexual desire and satisfaction.[19] If the alteration of our mental state through the use of drugs could lead to more sex, then could we not put together a plausible evolutionary hypothesis that suggests that, in some way, we may be 'willing hijackees' when it comes to taking at least some drugs? When it comes to alcohol, the most used and abused drug of all, the answer is probably yes.

We can construct an evolutionary argument for alcohol imbibition, and the evidence for it, by considering a number of different aspects of alcohol. First, it is easy to make. Fermentation, whereby yeast converts sugars to ethanol, happens naturally and as soon as we started growing crops like barley, wheat, sorghum or rice, then alcohol would have swiftly made an appearance. We didn't even need crops to make alcohol. Fruit can be gathered and fermented, while milk will ferment to produce *kumis*, a drink with a history that stretches back at least 7,000 years and an alcohol content of up to 2.5 per cent. The opportunity was there and history suggests we took it.

Second, alcohol makes us feel good and studies indicate that it has an effect on those all-important dopamine reward pathways of the brain.[20] We already know that we like to do things that make us feel good. Third, alcoholic drinks, at least in the early days, would have been nutritious and potentially safer to drink than the water available in some locations. Fourth, and most important, in moderate quantities alcohol is a powerful social lubricant. Drinking together would have acted to bond groups more closely, providing huge benefits for activities like hunting and for general social organisation. Of course, addiction to alcohol is a major global health problem and alcohol dependence is so great that sudden withdrawal can cause death, so there is a serious downside to it. However, in our earliest times, when we were adapting to alcohol intake, there weren't shops on every corner selling high-strength beer, wine and spirits. The alcohol levels, and the availability of alcohol in the environment in which we developed our drinking habits, made that environment very different from the one in which we find ourselves in the modern world.

So far so good, but none of these parts of the story necessarily imply evolutionary adaptation to drinking alcohol. Indeed, evidence that alcohol stimulates the reward pathway leads us rather neatly back to the hijack hypothesis, especially when paired with the nutritional benefits that consuming fermented beverages might confer. If we add in the value of social bonding and our ability to pass on cultural habits, then we need not invoke any further evolutionary argument. However, if we look further back into our evolutionary history we can come up with a very solid evolutionary argument for drinking alcohol. This argument has become known, rather charmingly, as the 'drunken monkey hypothesis'.

Central to the drunken monkey hypothesis is the fact that ripe and overripe fruits naturally ferment to produce alcohol. Robert Dudley, the proposer of the hypothesis, suggested that our early ancestors, and we are talking 10 million years ago, evolved a genetically based attraction (sensory and behavioural) to ethanol that allowed them to find these nutritionally rich prizes. The odour of ethanol, which carries in the air and could be detected at long range, would have allowed an animal attuned to it to home in on nutritious fruits in forests that might otherwise provide rather slim pickings.

The drunken monkey hypothesis finds support from the evolution of alcohol dehydrogenase (ADH) enzymes that allow us to breakdown the ethanol in alcoholic drinks. A study published in 2014 used the technique of paleogenetics to study the evolution and functioning of a particular type of ADH called ADH4. By looking at the evolution of genetic sequences, paleogenetics is able to 'resurrect' proteins from long-extinct ancestors. In this study, the researchers resurrected ADH4 from primate ancestors and showed that a major change happened around 10 million years ago. ADH4 is present in our mouths, oesophagus and stomachs and prior to 10 million years ago was rather poor at breaking down ethanol. The mutated version, on the other hand, is some 40 times better and equipped our ancestors for consuming fermented, alcohol-containing fruit. The timing is interesting too, since it coincides with our ancestors adapting to a life on the forest floor where ripe, overripe and fermented fruits would have been encountered. Also, at this time in Africa forests were shrinking and grasslands expanding. Avoiding open country, where our ancestors would have made an easy lunch for other creatures, would have been important, so an ability to find increasingly scarce fruit in the dwindling forests could have been a strong selection pressure on the evolution of ADH4. The old version of ADH4, incidentally, was very good at metabolising a different alcohol called geraniol as well as alcohols like cinnamyl, coniferyl and anisyl, which are produced by plants to prevent herbivores eating their leaves.[21] Our early, arboreal ancestors needed their version of ADH4 for leaf eating, while their descendants found the mutated version perfect for fruit.

The ethanol that ADH4 allowed our early ancestors to consume would have been present in very low concentrations and was, as a direct consequence of how it formed, consumed with food. The consumption of small amounts of alcohol in this way is known to have some health benefits. For example, considering all cancers combined, a study of middle-aged men found that cancer deaths were lower among those consuming up to one drink per day than they were in those who didn't drink at all.[22] Moderate alcohol intake can also protect against strokes that are caused by severely reduced blood flow to the brain (known as ischemic strokes).

The evidence suggests that we evolved an ability to consume alcohol very early in our lineage as a consequence of selection pressure on our diet. The ability to find and consume fermented fruits was an advantage and the small quantities of alcohol concerned may have conferred some health benefits. The interaction with dopamine reward pathways worked to link nutrient-rich and valuable food with pleasure. Much later in our evolution came the development of large-scale deliberate fermentation, following on from the development of agriculture. Deliberate fermentation created an environment of abundant and concentrated alcohol that super-stimulated our reward pathways. To some extent, the reproductive and survival advantages of social lubrication and group bonding that were provided by larger qualities of alcohol would have outstripped the health costs of increased alcohol consumption. Alcoholic drinks would also have travelled relatively well and provided nutrition. Despite these advantages, we were already taking steps towards the modern problems of alcohol abuse by changing our 'ethanol environment'. The modern world in which we now find ourselves is one in which alcohol is even more concentrated and easily available than ever before, and that makes for a very potent and dangerous evolutionary mismatch.

Risk addiction
So far we've dealt with substance abuse and developed a compelling case that our current problems around addiction and dependence are the consequence of a

hijacking-mediated mismatch between our evolutionary heritage and the modern world. Does this logic hold up for the other things to which we can become addicted, at least in the everyday usage of the word? It is worth noting that medical acceptance of certain behaviours as 'proper' addictions has been slow in developing, although recently behaviours including gambling disorder, internet gaming disorder, internet addiction, food addiction, hypersexuality ('sex addiction'), shopping addiction, exercise addiction and tanning addiction have become accepted alongside substance addictions. Addictions to sex and food (especially fatty and sugary food) make perfect sense from an evolutionary perspective. Sex is a primary stimulus for the mesolimbic reward pathway and if the opportunity is available to partake in a great deal of it, then it is easy to see how the reinforcement and learning mechanisms we have already met could take over. Food is also a primary stimulus for the reward pathway and fast food in particular provides the sort of calorie-dense feast that, at least in the short term, is highly rewarding. Pornography addiction, which is part of a wider issue concerning the use of pornography in an internet-enabled porn-rich environment, can also be explained through primary stimulation of the mesolimbic pathway, especially if coupled with masturbation and orgasm.

One widely recognised addiction that is of particular interest in an evolutionary sense is addiction to gambling. Often termed 'problem gambling', gambling addiction is now classified (by the *Diagnostic and Statistical Manual of Mental Disorders* 5th edition, or DSM-5) as an addiction and is properly called gambling disorder. Gambling disorder is gambling that causes problems either for the gambler (usually in terms of money and time) or for others (typically the gambler's family). A need to gamble, a preoccupation with gambling and the gambling of larger amounts of money ever more frequently leads to 'loss chasing' (where gamblers gamble harder after losing to recoup their losses), concealment of gambling behaviour, unsuccessful attempts to cut back or give up and symptoms of withdrawal (irritability and restlessness) when unable to gamble. Alcohol abuse is often a related issue and chronic gamblers may end up contemplating or attempting suicide.

Gambling is just that, a gamble, and while some knowledge and skill can mean that gamblers end up on top, the reality is that most do not. Take a look at a fruit machine and you'll see the payout percentage clearly displayed. The 'return to player' as it is called is usually between 80 per cent and 98 per cent, but is typically around 85 per cent. What this means is that if £1,000 is put into the machine the player will receive, on average, £850 back. This is a very poor return in anyone's book. Of course, you might get lucky and hit the jackpot, but you probably won't and the 'return to player' percentage says that over time you certainly won't; over time, you'll lose on average 15 per cent of whatever you feed in.[23] Slot machines are games of chance, but games that involve some level of skill (such as some casino card games) can improve the gambler's chances. Overall though, the odds are always strongly in favour of the house. Horse racing and other sports betting can improve the chances of winning for knowledgeable gamblers, but the continued occurrence of major sporting upsets confirms that there really is no such thing as a 'sure thing'. Given that gambling is a seemingly highly irrational and damaging behaviour, it is not surprising that a number of studies of gamblers have tried to construct a biochemical framework to understand why people do it and why they can't stop.

The results of these studies are not always that clear, with suggestions that some gamblers may gamble to increase levels of adrenaline, that gambling may be associated in some way with serotonin, that gambling might be linked with other impulse control disorders, anxiety disorders or personality disorders and, as we might now expect, that gambling causes stimulation of the mesolimbic pathway. One fact that is particularly illuminating when it comes to understanding how the mesolimbic reward pathway might be involved is that people experience a greater dopamine response if they don't know a reward is coming. The lights and sounds of slot machines especially are thought to enhance the 'surprise' of a big win and reinforce the 'high' caused by the unexpected dopamine surge.

Understanding the connections between gambling and reward is easier if we think of gambling more generically, as risky behaviour undertaken with the possibility of a large reward. In an ecological setting, taking risks can be highly beneficial in uncertain environments. Imagine for a moment that you are a mouse. All the time you are hiding in your mouse-hole somewhere, you are safe, but what you are not doing is eating or reproducing. Leaving your safe refuge leaves you open to predation, limiting survival, but gives you the opportunity to find food and mates. Understanding the trade-offs between risks and rewards, between costs and benefits, is fundamental to our understanding of the evolution of many behavioural and life-history traits in animals. Indeed, the emerging field of animal personality research, which studies the consistency of individuals' behaviour, places great emphasis on 'bold' and 'shy' as personality types and it is risk-taking behaviour that mostly defines these categories. If, under certain circumstances, risks are worth taking and risk-takers secure more resources and more offspring than risk-averse individuals, then it is easy to see how we could have evolved to reward such behaviours. Reinforce that risk-taking tendency with the bells, whistles and lights of the modern world and all its gambling opportunities, and firing up that reward pathway without taking life-threatening risks suddenly seems awfully tempting.

Understanding problem gambling from an evolutionary perspective opens up new treatments. Some suggest that compassion-focused therapies (CFT) provide an approach that allows addicts to understand better why they are in the situation so that they are better able to develop new strategies to manage their behaviour. As John Paulson of the University of Southern Indiana and author of 'Hardwired for Risk: The Clinical Utility of Exploring Evolutionary Aspects of Gambling', an article published in the *Journal of Gambling Issues* in 2018, put it, 'The hope is that coming to see and appreciate increasingly the influence of evolution on gambling will help those persons suffering to reduce their sense of shame and recover from their struggles.'[24]

Children in a modern-world sweetshop

Addiction of one form or another is one of the most damaging products of mismatches between our evolution and our current environment. By subverting our evolved reward system, highly destructive drugs and behaviours can become a significant factor in our lives and deaths. However, they can only do this to the extent that they do because of the great increase in 'potency' that the modern world provides. For drugs, potency is easy to understand but for other addictions it is no less the case. Fast food is both highly available and high in caloric value; a burger and chips is surely more potent than a basket of roots and berries. We can gamble on our phones, in pubs, in casinos, on computers and via the lottery and the combination of huge wins and accessibility means the gambling environment of the modern world is far more potent (although existentially less risky) than ever before. Overall, the modern world we have created interacts with our evolutionary tendency for chasing dopamine rewards, and the resulting mismatch leaves us as helpless as children in a sweetshop.

CHAPTER NINE

Fake News and False Beliefs

'Fake news' is everywhere. It's the rallying cry of Donald Trump, the curse of social media and the go-to excuse for just about anything that someone disagrees with. Fake news is simply any news that is presented as true but has no basis in fact, or is based on fact but contains significant falsehoods. Fake news is fabricated, made up, elaborated, embroidered, bogus, false, bullshit, or sometimes just simple barefaced lies. It is a problematic term. As the UK government pointed out when it banned 'fake news' from official documentation, it is 'poorly defined and misleading ... [and] conflates a variety of false information, from genuine error through to foreign interference in democratic processes'. Words like 'misinformation' or 'disinformation' are preferred in UK Government documents, and in cases where fake news is used to advance political causes or other agendas we might more correctly use the word propaganda. The near-constant accusations of 'fake news' by Donald Trump to dismiss more or less anything with which he disagrees has undoubtedly greatly weakened the term, or at least allied it with perceived political attacks by the 'liberal mainstream media'. *Washington Post* columnist and former public editor of the *New York Times* Margaret Sullivan has suggested that the term be 'retired', saying that 'When it started off, it actually meant what it sounds like – fraudulent or misinformation meant to deceive. It's now become kind of a punchline, something people use to poke fun. If you don't like something, you call it fake news.'[1] She clearly has a strong point but 'fake news' has nonetheless become a convenient if sweeping shorthand for different types of information that are in some way lacking in veracity. I am going to continue to use it throughout this chapter for reasons of convenience, but be mindful that it has its limitations and may already be a term, although not a concept, on the decline.

The crucial take-home point from the proliferation of fake news is that we seem very keen to believe things that aren't true. My argument will be that such misguided beliefs are manifestations of a series of mismatches between our evolutionary heritage and the modern media-drenched world that we have created. My aim is to explore the relationship between potentially evolved traits, or at least psychological components that may have been subject to selection and evolution (such as gullibility), and our ability to spot the truth in the modern world. There is an elephant in the room here though; since we are talking about 'belief' I could decide to drag religion into the argument. I am not going to go down that road and this decision warrants an explanation. I am an atheist, but I was brought up in what would be regarded as a Christian country in the late 1970s and early 1980s and so at primary school I was exposed to the Bible. At a very young age, listening to stories about Noah, or Adam and Eve, or Jesus working miracles and rising from the dead, I called 'bullshit'. Well, I didn't use that word but the gist was the same. None of it ever made any sense to me whatsoever, even as a five-year-old, although in later life I began to see behind the literal stories. The miracle of the loaves and fishes, so beloved by my primary school teachers that we chanted a song about it in assembly, is about getting people to share what little they have rather than some magical multiplication. And that is a good message. As I got older I found out about other religions and none of them ever moved me either. I know though that many people are moved, profoundly, by religion and that religion in many forms is an integral and important part of their lives. This chapter will not be yet another evolutionary biologist's attack on the 'stupidity' or otherwise of religion. Believing in religion is also not a modern phenomenon, but the massive influence, impact and reach of fake news is very much a feature of the modern world.

What I am concerned with in this chapter is whether our *ability* to believe false information is a product of a mismatch between our evolved cognitive characteristics and the modern world. If our ability to believe things, to be gullible to some extent, exists because of advantages afforded to us in our evolutionary past then it is not a great stretch to suggest that, awash in the sea of modern media, this ability has left us without the cognitive satnav we need to navigate our way through the lies. In other words, it is possible that evolution may have left us helpless in the face of our own deceptions, and that is an interesting if rather terrifying proposition.

Fake news is old news
Fake news might seem like a modern phenomenon, and certainly the ability to disseminate such stories quickly and globally is modern, but history is littered with

examples of what we might now term fake news. In the first century BC, Octavian orchestrated a fake news campaign, in this case perhaps better termed a propaganda war, directed against his one-time ally, but later bitter rival, Roman politician Mark Antony. Portraying him as a drunken womaniser and puppet of Queen Cleopatra, Octavian upped the ante when he got hold of a document he claimed was Mark Antony's official last will and testament. Whether this document was what Octavian claimed, and scholars are still debating whether it was real, fake or partially forged, reading it aloud in the Senate was a masterstroke. Legacies, including substantial pieces of Roman-held territory in the Eastern Mediterranean, were promised to the children Mark Antony had with Cleopatra. The will also claimed that the son of Julius Caesar and Cleopatra, Ptolemy XV Philopator Philometor Caesar (better known by the nickname Caesarion), was Caesar's rightful successor rather than Octavian, Caesar's adopted son. These and other damning statements played on the suspicions and prejudices Romans held against the East and against Cleopatra (an early example of confirmation bias, a topic we will return to later). The Senate in Rome stripped Mark Antony of his right to lead Roman armies and made him a traitor. The rest is history, in this case one written by the victor in the war Rome declared not against Mark Antony but against Cleopatra.[2]

The invention of the printing press allowed stories to reach far bigger audiences far more quickly than ever before and printed material purporting to be fact has long been a source of fake news. Such stories ranged from witches and sea monsters to more elaborate claims, like the assertion that the Lisbon earthquake of 1755 was divine retribution against sinners. This story spawned an entire new genre of fake news pamphlets in Portugal called *relações de sucessos*, attributing survival of the earthquake to an apparition of the Virgin Mary.[3] Another technological advance, the internet, allows a much greater platform for falsehood, reaching far beyond the 'World War 2 Bomber Found on the Moon' tabloid nonsense audiences are familiar with through publications like the *Daily Sport* or *The National Enquirer*. If you can't read 'Alien Bible Found! They Worship Oprah!', 'Saddam and Osama Adopt Shaved Ape Baby' (both classics from the publication *Weekly World News*, a largely fictional tabloid now available online only), or 'Statue of Elvis Found on the Moon' (part of the *Daily Sport*'s obsession with weird lunar objects) and call 'fake' then you might have a problem as an individual. The trouble with fake news in the modern world, though, is that it is no longer just a problem for individuals; the manipulation of the media through the dissemination of fake news has become a political matter and a problem at society level nationally and internationally. The evidence that the active dissemination of false information has swung elections and referendums ('All Aboard the Brexit Bus') leaves us facing a difficult truth: we are not good at distinguishing fact from fiction, and this has become a major issue.

Have we evolved to be trusting?

Believing fake news suggests gullibility and being gullible is widely seen as a negative characteristic. Gullibility though is fundamentally an issue of trust; a gullible person trusts that other people are telling them the truth, and without trust we are all doomed. If you don't believe me then imagine for a moment trying to live your life without trust. I am sitting in a house typing this on a laptop. I implicitly trust that the people who built my house knew what they were doing and that it won't fall down. I also trust the manufacturers of my laptop, since it is plugged into mains electricity and is sitting, literally, on my lap. A wiring fault could lead to a very nasty incident but I trust the safety markings on the plug, the power lead and the laptop itself. I trust that my wife isn't going to reach for a kitchen knife and kill me when she comes downstairs. In fact, despite our abilities and tendencies to be violent (see Chapter 7) I trust, every single day, that random people I meet won't kill me. Whenever I drive I trust that hundreds, or even thousands, of other drivers won't veer across the white line and crash into me. I trust that the guy who fitted my brakes didn't mess up the job deliberately or by accident. I trust that the factory making my pre-packed sandwich lunch followed food hygiene regulations and that I won't contract food poisoning. I trust that the vaccinations I insist on my children getting are what they say they are and not some government-sponsored mind-control serum. I even, despite occasional evidence to the contrary, trust that my children can play together unattended upstairs without causing too much damage to the house or each other. If you want any firmer evidence of the need for trust in society, pull out some money from your wallet. The grubby collection of worthless metal and paper, or whatever notes are made from these days, is testament to trust; you trust that someone will exchange goods or services for these trinkets and they in turn trust that a bank will take them and 'honour' them. Without trust we cannot have banks, and indeed a fundamental financial instrument is called a 'trust'. Trust, a firm belief in the reliability, truth or ability of someone or something, is everywhere, essential and, as we shall see, evolved.

There is a very thin line between the trust necessary for everyday life and the gullibility or credulity that can lead us towards believing fake news. In step with the importance we place on trust, and in particular the trust that we have in other people, trust has attracted a great deal of attention from psychologists and biologists. The nuances of trust, the levels of trust and the manipulation of trust are subjects of ongoing socio-psychological research, and such studies are revealing some fascinating insights at a biological level. A particularly interesting study was published in 2005 in the journal *Nature* and reported the effects on trust of giving people oxytocin. Oxytocin is both a hormone and a neuropeptide, a substance used by neurons to communicate with each other. We met oxytocin in Chapter 6 where we explored its role in social bonding. Levels of oxytocin increased in individuals who were gossiping and so increased their level of bonding, which feels like a potentially important step towards building trust. Oxytocin is also postulated to be involved in perhaps the most fundamental of all social bonding, the bond between a mother and infant.[4] By locating oxytocin receptors in the brain we can begin to get more of an idea of what functions oxytocin serves, and doing this points us yet further towards a possible role in developing trust. Studies of non-human mammals reveal that oxytocin receptors are located in regions of the brain involved in pair-bonding, maternal care, sexual behaviour and the ability to form social attachments. The implication of this receptor mapping is that oxytocin may be involved in allowing animals to overcome a natural avoidance of physical proximity and enable what is called 'approach behaviour'. The authors of the 2005 *Nature* paper went a step further and hypothesised that oxytocin might be involved in promoting 'prosocial behaviour' in humans, specifically trust.[5]

To investigate the role that oxytocin might play in developing or maintaining trust, the experimenters set up a game based on financial investment. Two subjects played the game and adopted the role of either investor or trustee. Although the game was played with 'monetary units' (MUs) these were exchanged into real money at the end of the game (a whopping 0.4 Swiss Francs per MU), so there was a real-world incentive to do well. Both parties start with 12 MUs and to get things rolling the investor has a decision to make: how much money to transfer to the trustee? This is where things get interesting because the experimenter triples that amount before telling the trustee what was transferred. So, let's say the investor decides to go 'all in' and transfer 12MUs to the trustee. The trustee is told that they now have 48MUs (their original 12MUs and the tripled transfer from the investor). The trustee now has a decision to make: how much to 'back transfer' to the investor? The back transfer isn't tripled but can be anything between 0MUs (not so good for the investor) or their full balance (48MUs and not so good for the trustee). The trustee can honour the investor's trust by sharing the money gained such that both players end up better off, or the trustee can violate the investor's trust and simply pocket the profit; they don't even need to pay off the investor's 12MUs.

A rather dry experimental system perhaps, the trust game was spiced up by the prospect of actual money (a rare commodity in scientific experiments), the jeopardy of distrust and the application of oxytocin to half the players via a nasal spray. The other players received a placebo nasal spray that contained all the same ingredients but was lacking oxytocin. The hypothesis that oxytocin increases trust makes some very simple predictions, the first of which is that investors that have received three puffs of oxytocin up the nose should be more trusting (give up more of their 12MUs) than those that have been inhaling a placebo. If, on the other hand, oxytocin makes no difference to trust then we should expect no difference in initial investment between the oxytocin and the placebo group. In fact, what they found was that 13 of the 29 investors (45 per cent) who took oxytocin showed maximum trust (going all-in and investing 12MUs) compared with only 6 of the 29 (21 per cent) in the placebo group. Low levels of trust, defined as transfers below 8MUs, were relatively common in the placebo group, accounting for 45 per cent of transfers, but much rarer in the oxytocin group, accounting for just 21 per cent. On average, oxytocin takers had 17 per cent higher transfers overall, with a median amount (the 'middle' value if all the amounts were lined up in order) transferred of 10MUs compared with 8MUs in the placebo group. Interestingly, even the placebo group seemed pretty trusting, giving up two-thirds of their money. This perhaps gives us some insight into how people end up falling for get-rich-quick schemes.

There was a slight complication in all this because we could interpret the results of administering oxytocin not in terms of increasing trust but as a lowering of the threshold for risky behaviour; oxytocin makes you more likely to go all-in because, you know, why not? To investigate this possibility the experimenters ran the same basic experiment, but instead of there being a person acting as a trustee and making decisions, the back transfer was determined at random based on the back transfers that took place during the first experiment. So, investors, on average, had the same level of risk but there was no social interaction with a trustee and so no need for interpersonal trust. In this experiment, the oxytocin and the placebo groups did not differ and both groups were actually the same as the placebo group in the first experiment. Only the oxytocin group in the first experiment, where interpersonal trust was required, was different in any way and that difference was that they were more trusting.

This seemingly clear link between oxytocin and trust caused, predictably, a big stir when it first came out. It led to oxytocin being dubbed the 'trust molecule' and it spawned several follow-up studies using the same method. It was at this point that the problems emerged. Some follow-up studies failed to replicate the effect of inhaling oxytocin on trust in similar games, and other studies, using questionnaire approaches, also drew a blank. In 2015 a major review of the link between oxytocin and trust looked at three lines of evidence: the trust experiments; individual oxytocin levels in blood and levels of trust in that individual; and genetic differences in the oxytocin receptor gene that might be associated with trust. Sadly, for those of us who enjoy a good story, their conclusions were that 'the cumulative evidence does not provide robust convergent evidence that human trust is reliably associated with OT [oxytocin] (or caused by it)'.[6]

So much for the simple story. However, recent work is showing that oxytocin does seem to have a role to play in developing trust, but that role is complex, context-dependent and likely involves learning. Scanning the brains of male subjects undertaking tasks within an fMRI (functional Magnetic Resonance Imaging) machine, which measures and maps brain activity, allows researchers to see what parts of the brain are active at any given time.[7] These scans have shown that oxytocin, rather than promoting general feelings of generosity that might induce trust, actually seems to reduce 'feedback learning' by subduing communication between parts of the reward circuitry in the brain that we met in Chapter 8 when we explored addiction. The result of this reduction in learning is that, under the effects of oxytocin, we are more likely to go with our underlying beliefs, rather than make more nuanced decisions. As the authors conclude, oxytocin alters 'the brain's encoding of prediction error and therefore its ability to modulate pre-existing beliefs'. This permits our innate biases (whether positive or negative) towards a new social situation to dominate, which could promote more trusting behaviour.

Trust, despite some problems replicating the early research, does appear to have a biochemical basis in the brain. As we might expect with such a complicated concept, that biochemical basis is equally complex and involves a number of aspects of the brain including learning and the reward circuitry. The evolution of trust is also likely to be complex but nonetheless we can, with some confidence, suggest that our ability to trust is an evolved characteristic. A study published in 2008 measured identical and non-identical twins' performance in a trust game, and the results strongly indicated that 'humans are endowed with genetic variation that can partially account for differences in trust and trustworthiness'. Trust can be seen as cooperative behaviour while being distrustful is non-cooperative, and the authors of the twin study tentatively concluded that the existence of genetic variation in levels of trust lends support to the idea that cooperating and non-cooperating behaviours could co-exist in humans. The selection pressures for trusting behaviour are likely to be based on the advantages provided by cooperating and working together, highly social behaviour that requires individuals to trust each other. On the other hand, there are situations where personal advantage might be gained by being distrusting. The fact that trust seems to be linked to our reward circuits and to learning, and is generally context-dependent (we don't simply have a universal 'trust level' but can vary trust depending on the situation and the individuals with which we interact), indicates a complex evolutionary history tightly interwoven with our developing social networks (Chapter 6). Perhaps it was actually the evolution of trust that allowed our ancestors to develop more sophisticated social interactions and to make full use of the power that larger groups can provide? As a simple example, it is hard to imagine hunting a large animal in a group without trusting the person holding a sharp spear and standing right next to you.

I am all too aware here that I might be straying in to 'Just So' territory with some hand-waving about putative selection for and evolution of trust. The potential biochemical nature of trust, twin studies showing heritability and the importance of trust in so many aspects of our lives, certainly lend support to an evolutionary history though, and further evidence comes from theoretical approaches. Trust and cooperation have been studied using the modelling approach called game theory. Game theory allows us to consider the interactions between individuals and has proved to be a powerful and useful approach in understanding more complex aspects of social behaviour and evolution.[8] The models produced by game theory studies have shown that trust and trustworthiness can evolve, and real-world factors can result in the sort of variations of trust that we see in humans. We are still very much at the early stages of learning about the origins of human trust, but recent work on chimpanzees suggests that comparison with non-human primates and other social animals may yield valuable insights. One such study presented chimpanzees with a version of the human trust game and showed that individuals trusted their friends (determined by observing the chimps and working out who hung out with whom) more than their non-friends.[9] Overall, I think we are on firm ground to suggest that trust is an evolved characteristic in humans, albeit one that is complex.

Misplaced trust

Believing fake news is an example of misplaced trust. Until the rise of the internet, and more especially the rise and domination of social media and other platforms for disseminating news, we got most of our information about the wider world around us through printed newspapers and limited TV and radio news. We may also have got some of our information through our friends (trusted sources), who may have read or seen something interesting and passed it on to us face to face. Organisations like the BBC were trusted and information passed on by them was considered 'real news'. The political affiliations of many newspapers were clear but there was rarely the thought that they were actively lying to us. More importantly though, the ways in which we received news were limited and to a certain extent austere; it was something of an event to sit down and watch 'the news'. Nowadays this is not the case. The internet may have compromised the financial security of traditional physical newspapers, but it has created a truly dizzying splendour of online news channels. Meanwhile, cable TV, deregulation and widespread acceptance of advertising have driven up the amount of news coverage on our TV screens from a scheduled short bulletin being read every so often by a dour man in a suit to the glitz and glamour of 24 hours a day, seven days a week rolling coverage across multiple channels. Our social media feeds are stuffed full with links to online news stories and current affairs videos. In short, we are inundated with information coming at us from all directions in a complex and high-volume information stream. In no particular order we have state-sponsored TV, online news channels, old channels, new channels, satirical news, commentary, review, breaking news, ongoing news, throwback news, dense analysis, superficial analysis, zero analysis, space fillers, advertorials, editorials, 'conspir-atorials', video, text, cartoons, infographics and interactives. Just as we saw in Chapter 8 with drugs and alcohol, the modern world is one of greatly enhanced potency and availability.

Simply trusting everything we read or is told to us by someone who 'looks the part', like a newsreader, fits the definition of gullibility. Being gullible, being too trusting of information given to you, is a quality often associated with children. A three-year-old will look at you wide-eyed as you tell them about Santa Claus or the tooth fairy because you are their parent and you have formed a trusted bond; they believe you. This is handy for evolutionary success because children who do the exact opposite of what their parents tell them are far less likely to make it to adulthood. Children have an awful lot to learn in a very short timeframe and believing what they are told, being gullible, is a useful time-saver. However, believing what your parents tell you might be a good survival strategy initially, but at some point there needs to be a degree of criticality because gullibility can only take you so far. Studies of children and their relationships with true and false information give us interesting insights into how rapidly we develop into less gullible and more critical thinkers as we age, and can also throw some light on why we fall for fake news.

There are several ways that a child can assess the accuracy of information they are being given by someone. One way is to assess the accuracy of previous information given by that person. If someone seems to have been accurate and told the truth in the past then it seems reasonable to assume that they are being accurate in the present. A wealth of studies of children show that they do indeed prefer to learn from those who have previously given accurate information, rather than from those who have previously been inaccurate.[10] This is of course equally valid for adults; if someone or something, like a TV news outlet, has been accurate in the past why doubt it now?

With TV news especially, there may also be a strong 'halo effect' at play. Halo effects describe the tendency for impressions created in one area to positively influence opinion in another area. We tend to package certain traits together; for example, we judge more attractive people to have more socially desirable traits like intelligence,[11] altruism and, crucially, trustworthiness. Studies have repeatedly shown that we also prejudge attractive people to be 'better', to hold more prestigious and secure jobs and to have happier lives. The cognitive bias inherent in the halo effect might at first glance, and even second glance, seem irrational; how can we possibly judge unseen characteristics of an individual simply by looking at their face? Once again, our evolutionary heritage might provide an answer.

In many animal species one sex, typically males, compete for the attention of the 'choosy' sex, typically females. The reason why it is often males that compete and females that choose is linked to the large investment that females put into producing their reproductive cells (eggs) versus the relatively smaller investment males put into sperm. This investment differential can be further enhanced by females providing some level of parental care beyond that provided by males. Enhanced female care can begin at the very start, by providing a rich yolk to the egg (for example in birds, reptiles, amphibians, fish) or by nurturing the young within their own bodies (as is the case in mammals). The upshot of this increased female investment and parental care is that while males are nearly always ready, willing and able to provide copious amount of sperm, females are not necessarily available to receive it; eggs take time to produce, or they may be pregnant, sitting on eggs, or lactating young. In a population with an equal sex ratio this imbalance leads to more males being available than there are reproductively receptive females. The imbalance in what is called the operational sex ratio and the consequent surfeit of males allows females to be choosy. Females that pick the 'best' males, healthy males with 'good genes' that will lead to lots of healthy fecund offspring, will do better than those that are less choosy. In species where males provide the larger investment, such as seahorses and pipefish, where males keep eggs and hatchlings safe in a brood pouch accessed through an opening on their abdomen, it is males that will be in short supply. In these more unusual 'sex role-reversed' species it is females that compete for males. This 'competition/choice' process leads to sexual selection whereby one sex can end up with adornments and weapons that make them better able to compete with other males (antlers, horns, strength), or more attractive to females (peacocks' tails, bright colouration, crests), or both.

One aspect of sexual selection that is potentially relevant to the halo effect is the concept that has become known as 'fluctuating asymmetry'. We are symmetrical animals with our left side and right side more or less identical, at least externally, through a line of symmetry that runs between our eyes and passes through the groin. Studies have shown that more symmetrical faces are judged to be more attractive and this is the basic principle behind the 'How Attractive Am I?' clickbait that haunts Facebook. We are not, however, perfectly symmetrical, and deviations from the 'ideal' of symmetry creep in during our development as different genetic and environmental pressures are exerted. Symmetry therefore might indicate 'developmental stability', and be reflective of overall genetic quality; 'better' individuals are more likely to be symmetrical. In animal studies, including insects, birds and mammals, symmetry of traits (where structures on the right side are compared with those on the left) has been shown to correlate with mating success; symmetrical males are picked more by females. In human studies it seems that symmetrical faces are not just a feature that we rate as more attractive. There are studies that link facial asymmetry with lower intelligence, lower levels of extraversion (how outgoing and sociable people are) and more intellectual impairment with ageing. One study even found that criminal offenders had lower facial symmetry than non-offenders.[12] Delving deeper, studies have found links between lower levels of facial symmetry and a greater vulnerability to parasites, as well as lower levels of disease immunity. Correlations have also been found with mental health, with lower

symmetry being associated in one study with higher
levels of depression in men. Several studies have
associated lower symmetry with schizophrenia. For
obvious reasons such studies are often controversial
both socially and scientifically. A recent review of the
literature in this area concluded that while some
associations, such as with schizophrenia, are repeatable
and robust, other studies 'have not always followed best
practices' when it comes to measuring features and
determining the tiny differences involved.[13] Nonetheless,
the balance of evidence allows us to say that symmetrical
faces are judged as more attractive and that facial
symmetry is potentially indicative of a wealth of
underlying characteristics that relate to possible 'mate
quality' Picking a good partner is an important decision
for humans, and when viewed through an evolutionary
lens the halo effect is not so irrational. So, the next time
you watch an attractive, well-groomed newsreader, be
mindful of your cognitive biases; the halo effect means
you are more likely to think of them as trustworthy and
to believe whatever they are saying.

Halo effects also apply to non-living entities including
corporate brands, as companies like Apple and McDonalds
know all too well. This may be be a factor in news output
too. Glossy, well-produced and generally factually
accurate news programmes with all the accoutrements of
the modern news format, such as infographics, outside
broadcasts and in-studio punditry, all conspire to create a
halo effect for any similar output. Even if the information
is inaccurate, the corporate halo effect lures us in and
the individual halo effect makes us want to trust the
newscaster.

We believe authority

The idea that fake news is misplaced trust is strengthened by examining the relationship between trust and authority figures, including doctors, scientists and politicians. Our tendency to obey authority figures is most commonly illustrated by reference to the famous Millgram experiment. In these experiments of the 1960s Stanley Millgram instructed participants to administer what they believed could, at times, be fatal electric shocks to another participant (actually an actor) under the guidance of the 'authority figure' (the experimenter). Criticism of interpretations of the experiment have led to a different framing, moving from 'blind obedience' to 'engaged followership' where 'people are prepared to harm others because they identify with their leaders' cause and believe their actions to be virtuous'.[14] In fact, recent work further re-frames the study and suggests that participants sought to justify their actions not through some sense of virtuousness but because they doubted that anyone had really been harmed.[15] Other more subtle studies have been carried out to demonstrate what has become known as the authority principle though and the findings are consistent; we tend to trust those that we regard as being 'authority' figures. Here again the halo effect rears its unwelcome head, and we tend to regard the opinions of authority figures as more reliable than our own even if they are expressing an opinion on matters outside their sphere of expertise. The traditional authority figures of such experiments are doctors, lawyers, scientists or law-enforcement officers, but such implicit trust in those we regard as being 'in charge' or 'expert' translates very easily to actors and other celebrities and can certainly be applied to newsreaders, media commentators and others involved in conveying information through the media.

If you're going to lie, be confident

Another source of information available to us when deciding whether we are being told the truth is confidence. Studies of children show that we develop some skills for gauging confidence early. Three- and four-year-olds are aware of the differences between 'guessing' and 'knowing' and are more likely to believe information given to them by someone who 'knows' rather than by someone who is expressing a degree of uncertainty. The ability to spot confidence and associate it with knowledge actually starts much earlier in life, before children have firm verbal skills. Two-year-olds have been shown to be aware of, and sensitive to, non-verbal indications of confidence. These younger children are more likely to imitate someone's actions if that person has expressed those actions with confident body language. What is interesting is that as children get to the age of four they start to be able to make judgement calls based on a combination of accuracy and confidence. They prefer to believe information given to them by a confident individual, but if that information turns out to be inaccurate they start to prefer hesitant but previously accurate individuals.

An evolutionary mismatch

What is starting to emerge are the perfect evolved intellectual conditions for the modern world of fake news to take root. First, as a social species we have evolved trust and most of us tend towards being trusting unless evidence suggests otherwise. Remember, most investors in the trust game were happy enough to give away two-thirds of their money. Second, as a species that is essentially helpless for much of our early lives, we have neural mechanisms that even before we can talk allow us to interpret the 'truthfulness' of those around us. Innate neural mechanics, gradually honed through experience, lead us towards believing those who present information confidently. Halo effects, perhaps based on deep-rooted selective advantages linked to mate choice, cause us to be dazzled by certain individuals and corporations, which of course tend to present themselves towards the 'extreme' end of the confidence spectrum. We also seem to have an innate, and highly sensible, tendency to believe sources of information if they have previously been shown to be accurate. When we pull all these factors together and hose them down with a high-volume stream of information originating from previously trusted, confident and attractive sources, then it is little surprise that we fall for falsehoods. Mix in a deliberate intention to deceive, perhaps politically motivated, and then we are two-year-olds on a play mat willing to follow the grown-up with the most confident body language. All that is before we consider another set of evolved tendencies that have an important part to play in the story of fake news.

The importance of 'belonging'
Our cognitive biases towards confidence, previous accuracy and, through halo effects, attractiveness are important in our evolved predisposition towards believing

fake news, but are only part of the story. There is another powerful part of our evolutionary heritage that makes us particularly vulnerable to fake news and perhaps to other seemingly incredible sources of information: group identity and confirmation bias. Social identity theory was developed in the 1970s and 1980s by social psychologists interested in understanding intergroup behaviour. It is a big topic, has been the subject of many papers, articles and books and has not been without some controversy. At its heart though is a simple concept that has survived the test of time and rigorous debate: our behaviour is often a compromise, balancing interpersonal and intergroup interactions.

We identify both as individuals and as members of various groups. That group might be a close family group, an extended family, your school, your university, a football team, a political party, a religion, a nation and so on. By belonging to that group, and identifying as part of it, you become part of the 'in-group' and are likely to display in-group biases. These biases mean that you will give preferential treatment to others in your group while also potentially biasing against 'out-groups'. An important aspect of social identity theory is that members of the in-group will seek out negative aspects of the out-group and in so doing bolster the in-group's self-esteem.

Social identify theory is a concept sometimes summed up by the word 'tribalism'. Although this simplifies the issue, and brings in unhelpful or even offensive connotations, it is nonetheless useful shorthand for a concept that has proved its worth in social psychology. It is not difficult at all to see why we might have evolved strong tendencies for behaviours that reinforce whatever group we are in. As a social species we rely more than we often think on interpersonal interactions and, certainly at times in our evolutionary history, intergroup conflict would have been a real and important selection pressure. Group identity and enhanced trust in the members of that group will have been important components of group, and thereby individual, success. So, how does group identification fit in with fake news? Well, the tendency to be biased positively towards those we perceive as being in our group can extend to being biased towards information that fits our preconceptions about our group and others. This is likely to be an especially powerful cognitive bias when the information being presented is positive about people in our group (the in-group) or negative about those in out-groups. Being swayed towards what we already think is the basis of another big issue with fake news: confirmation bias.

Fake news preys upon confirmation bias, potentially deliberately if falsehoods are being peddled as part of a wider agenda. Confirmation bias is a serious problem and it affects how we search for information, how we interpret information that we receive and how we remember that evidence later. Bias in the way that we think is not a modern phenomenon. Dante Alighieri defines it beautifully in the early fourteenth century in the *Divine Comedy* when Thomas Aquinas meets him in Paradise and says 'opinion – hasty – often can incline to the wrong side, and then affection for one's own opinion binds, confines the mind'. Understanding the precise neurological mechanisms that underpin our thinking, and our biases, is one of the great works-in-progress of modern biology but it isn't too difficult to frame some plausible scenarios that would support an evolutionary, and therefore heritable, basis for confirmation bias. Strengthening group membership and enhancing social identity and cohesion could certainly provide a strong selective advantage for biased thinking processes, provided that the costs did not exceed the benefits. Thought processes are time-consuming and shortcuts that allow our brains to process information rapidly could be favoured, again if the costs are small compared to the benefits. Such shortcuts use common-sense rules of thumb and are known as heuristics. One suggestion is that we aren't so much interested in the absolute truth of a matter but rather we seek to avoid the most costly errors. Let's imagine two friends who trust each other, as friends often do. One friend suddenly has the thought that the other is dishonest or untrustworthy. The suspicious friend now has a choice: either question the integrity of the other person and erode the friendship (potentially very costly), or seek evidence that confirms the friend's honesty. If it is less costly to go along with the status quo (and friends are often hard to come by), then it makes sense to be biased because it could cost less than destroying the alliance.

We can see the importance of both confirmation bias and group identity in the acceptance and spread of fake news on social media threads. Climate change is an excellent example. The evidence that climate change is happening and that it is anthropogenic in origin is overwhelming. That Earth's climate is warming is accepted as fact by more than 97 per cent of actively publishing climate scientists. Despite this consensus and the weight, strength and diversity of evidence supporting it, climate sceptics still abound. When a post appears that purports to show that climate change is not happening, or when the President of the United States suggests it is a hoax perpetrated by China, you can guarantee that this 'news' will be posted gleefully by climate sceptics because it confirms their bias. Reading the comments arising from these posts, as I often do, confirms the in-group/out-group nature of these debates, which play out very much as a 'them versus us' conflict. Typically, such conversations only serve to polarise the groups further, leading to entrenchment and even more aggressive sharing of the fake news. Eager to support their argument further, people will search out yet more confirmatory evidence for their point of view. You can even see comments between friends within these threads referring to 'secret', private groups where more confirmatory news is being shared. These exclusive 'echo chambers' further feed the bias. Much of this activity perhaps stems from our evolved tendencies towards trust and reinforcing group identity, and from cognitive biases that develop from useful heuristics that are usually beneficial, or at least cost-reducing.

Our ability to fall for misinformation is not new, but the modern world, with its enhanced potency and availability, provides so much more opportunity for our evolved vulnerabilities to be exploited. We are not going to evolve our way towards increasing criticality, but we can learn to adapt. With an increasing awareness that not all we read is true, and an increasing exposure to the issues that fake news presents, we can perhaps be optimistic that in the future we may be more discerning information consumers. This is especially pertinent for those who are growing up as 'digital natives', with no memory or experience of a world before the internet, social media and modern fake news. A greater understanding of why we believe misinformation, built perhaps on an understanding of the cognitive mechanisms we have evolved to facilitate social living, is going to be hugely important in the future if we are to function in the modern digital age. But what does 'the future' really mean to us and what might it hold for an upright ape increasingly finding itself unfit for purpose? These are the questions that I will try to answer in the final chapter.

CHAPTER TEN

The Future

Our evolutionary heritage, extending from the deep roots of our phylogenetic history to more recent genetic changes, for example in response to agriculture, provides the fundamental basis for our ability to deal with whatever complications the world throws at us. The anthropogenic nature of much of the modern world means that we are often throwing complications at ourselves and, as we have seen, our evolutionary heritage is not always able to cope. In fact in many cases our heritage is working against us, leaving us vulnerable to all manner of problems in the world we have created. These problems, ranging from our diet to addictions and from violence to fake news can be explained in part by the mismatch that has developed between the modern world and the environment in which we evolved.

Of course, we are far more than simply the product of our genes and it is the interplay between our genetic 'nature' and the 'nurture' of the environment, as well as our unparalleled ability to learn and adapt mentally, that defines us and allows us to function. The current human population is a testament to the fact that despite being 'unfit for purpose' in so many ways, we nonetheless manage to do more than just scrape by. Our ability to shape and construct our own environment makes us the greatest of all of nature's engineers but, unlike the nutrient cycling achieved by termites in their impressive mounds or the European beaver's creation of biodiverse flooded forests, our environmental engineering mostly suits just one species: us.

One could argue that the greatest example of our 'unfitness' in the modern world is that the evolution of a brain large and complex enough for abstract thought and social cooperation, coupled with our unusually dextrous hands, have provided the ideal tools to engineer our own destruction. The past 75 years, just three generations from an evolutionary perspective, have seen the rise of cheap air travel, widespread car ownership, 'consumerism', the internet, real-time globalisation and a tripling of the human population. We are degrading our own habitat, changing the climate and generally dirtying our own nest, and the solutions are not going to come from some miraculous set of mutations and evolutionary changes. Even if it were possible to evolve our way out of the environmental mess we are rapidly making, and let's be honest, it isn't possible, we simply don't have the time. If we are to avoid an environmental disaster of our own making then we going to have to use the most powerful and important evolutionary legacy of all, and the reason we are in this mess in the first place: our brain. The problem is that when it comes to planning for the future, evolution has a lined up a couple of other serious mismatches that we are going to have to negotiate.

Natural selection: selfish and immediate
We are in the unique position of knowing that we are heading for existential trouble, knowing why and being

able to do something about it. No animal has ever before had this luxury. Dinosaurs roaming the Earth at the end of the Cretaceous didn't look up to see an asteroid approaching and decide to make a plan. I undertake research on a fenced wildlife reserve in South Africa. The antelope grazing there on the rich grassland don't have the wherewithal to plan and enact a regime of rotational burning to keep the grass capable of sustaining them into the future. Without management these animals would eventually eat all the grass and, being unable to move on to pastures new, would starve. This has actually happened on a number of reserves that fail to manage habitat and stocking densities correctly, and it is a good analogy for what the future on Earth might be for us. Unlike wildlife, humans can make plans and we can manage resources if we choose to. At this moment, which many are suggesting is a crucial one in terms of our long-term survival, making a plan and managing our resources are precisely the things we need to do. This will require us to think beyond our own limited self-interests and plan far into the future, two characteristics that evolution has tended to work against.

Natural selection and evolution are all about the 'here and now' and the immediate future. What behaviour will gain you a mate? How can you avoid being eaten? What maximises offspring production? Evolution is not striving towards a grand plan, some relentless quest towards the perfect future solution. The idea of evolution working in some goal-directed way is called teleology and teleological thinking is very easy to slip into if you're not careful. Giraffes don't have long necks because they 'want' to reach leaves high in the canopy. Rather, giraffes evolved long necks because across generations individuals at the longer end of the spectrum of neck length, which had some heritable component, were more successful than those with shorter necks. At some point the costs of extreme height and a long neck (the ever more robust mechanical support required and the high blood pressure necessary to supply the brain) exceeded the benefit these traits provided and put the brakes on the evolution of neck length. At no point during the process were giraffes, or evolution, planning for the benefits of future generations.

While we rightfully pride ourselves on our intellectual abilities, our brain is simply another adaptation that enables us to leave more offspring. This inevitably tends to make the brain a tool for the here and now. We are very good at attending to those issues which are immediate, like all animals, because it is the 'present' that brings challenges that we have to overcome. That is not to say we cannot make plans for the future; clearly we can and do. But it does suggest that intellectually we might have a different relationship with the abstract future than we do with the concrete present. Other animals can also make plans for the future; squirrels bury nuts for themselves for winter, while some birds store seeds in holes in tree trunks that they will find (using their remarkable spatial memory) and eat later. What we don't see in nature though are squirrels or birds intentionally storing food for future generations. Of course, buried nuts that are forgotten may germinate and produce trees that could feed future generations but a lucky accident is hardly future planning. Nut-burying behaviour and the cognitive structures required to find them again have been selected because of advantages accruing to the individual doing the burying in the short-term future. The very real present is always more important than the abstract future, and it seems more than reasonable to suggest that thought processes that can get an individual out of trouble right now, and that originate from neurological arrangements with a genetic basis, will flourish. We will return to the issue of our relationship with 'future us' shortly, but for now we also need to consider our relationship with everyone else, because as we will see, this has an interesting part to play in considering how we think about the future.

Self-interest is hard to ignore
Given that selection acts on individuals, we should have a strong tendency to value ourselves above others. Certainly self-interest is the rule in most of nature, and we successfully make sense of the evolution of much of animal behaviour by viewing it through the lens of 'selfishness' and individual benefit. Of course, there are plenty of humans who seem to act selflessly, who give their lives for others, who devote their time to helping others, who give away their fortunes and so on. Equally, there are also plenty of people who aren't violent (Chapter 7), don't get addicted to substances (Chapter 8), can function entirely healthily on social media (Chapter 6) and have a wonderful gut flora (Chapter 4). The key word here, as indeed has been the case throughout this book, is *tendency*; the question we are really asking is, do we have a *tendency* to value our own interests over those

316 UNFIT FOR PURPOSE

of others? If we accept that some of our thinking processes are the hard-wired consequence of evolution with the chance of being further honed by experience and learning, and that selection acts predominantly on individuals, then thought processes that tend to support self-interest seem more than likely. But as we developed into a more social species, and the benefits of being social began to emerge, then selection would also seem likely to favour intellectual processes and outlooks that were more favourable towards our social group (very often our kin). It is this balance between 'self' and 'group', or 'others', that is crucial.

Even in groups we often think of as being solely about 'group interest' we can still see the signature of self-interest. Honeybees are often portrayed as the ultimate in harmonious group living. The colony is headed by a single queen who lays all the eggs. She is the 'breeder', the 'reproductive', and the female workers that emerge from these eggs are sterile (non-reproductive). The workers toil ceaselessly to grow the colony until at some point it is large enough to produce a swarm. A large group of workers (about a third of the total number usually) departs with the old queen while a new queen is reared inside. It is the colony itself that is reproducing (we start with one colony and a swarm leaves us with two) and it is the colony that must get large enough to be able to produce a swarm. The queen–worker division of labour, with queens laying eggs and workers doing all the other work, is efficient and effective but it implies that workers can only benefit from what is called 'indirect fitness', by helping to rear a sister queen who will head a new colony of their nieces.

But the reality is that worker bees, and indeed the workers in many ant, social bee and social wasp colonies, are not in fact completely sterile and can gain some 'direct fitness' by having offspring. They can lay unfertilised eggs, which through a process of sex determination called haplodiploidy will hatch into males. If the queen has mated more than twice then the relatedness within the colony drops to a point where the workers value other worker-laid sons (their nephews) less than their own sons and less than their brothers (the queen's sons). In this case, even though any given worker wants to lay eggs, the worker collective will seek to destroy eggs laid by their sister workers. This so-called worker policing can be highly costly, wasting resources and time, but self-interest can nonetheless allow for worker egg-laying to persist. That such self-interest is evident, even in much more group-oriented species than we will ever be, shows what a very important role it has.

There's room for hope
Given the importance of self-interest and cooperation in our everyday lives, the intellectual interest that we clearly have in our own behaviour and the links to wider issues such as morality, it is no surprise that psychologists have done their best to answer the question 'are we selfish?' The overall conclusion is put rather nicely by Emily Pronin and colleagues in a 2008 paper that we will return to shortly and that is largely concerned with investigating how we think about 'future us'. They put it like this: we have a 'tendency to make decisions that neglect others relative to the self'.[1] A guarded, carefully

worded if damning statement. However, recent work suggests we could be less instinctively selfish than we might think.

In a paper entitled 'Spontaneous giving and calculated greed' published in *Nature* in 2012, Joshua Greene and colleagues set up an experiment whereby they could investigate the effect of 'thinking' on our behaviour.[2] The idea underpinning the research is that we might act more selfishly, displaying what they term 'rational self-interest', when we get a chance to think about our actions, and less selfishly if we act under a time restraint such that we can't think too much about what we are doing. On the other hand, as they put it, perhaps we are simply 'predisposed towards selfishness, behaving cooperatively only through active self-control'. In other words, some time to think might instead make us better people, less inclined towards self-interest. Their experiment allowed them to distinguish between these two outcomes and to determine the contribution made by intuition and reflection on our selfish/selfless behaviour.

The experimental set-up was very similar to the investor-trustee game we met in Chapter 9 when we considered the role of oxytocin in developing trust. Participants were assigned roles, given money to invest and once again the experimenters multiplied the payment (this time doubling rather than tripling the amount invested) before dividing up the spoils depending on decisions made by the trustees. The interesting variation added by Greene and colleagues was to introduce the element of time; in some of the games the participants had to make their decisions rapidly (within 10 seconds), but in others they were forced to take time to reflect on their decision. Rapidly made decisions are more likely to be the product of 'intuition', evolved heuristics that help us in the here and now, whereas decisions that come from reflection are likely influenced by higher-level intellectual processing. The results across all the experiments showed that when participants were forced to make a decision rapidly they were more cooperative than when they were forced to reflect on their actions. The overall conclusion was that cooperation is intuitive. They suggest that this is because of 'cooperative heuristics', simple cognitive rules of thumb, which kick in when we need to make rapid decisions. Overall, cooperation is typically advantageous and so such a rule of thumb mostly works out for us. Reflection, the time to think about a decision, on the other hand tends to lead us away from cooperation to act more selfishly.

The 'social heuristic hypothesis' provides a way to interpret these findings by using, as we would now expect, a combination of nature, nurture and learning. The hypothesis suggests that individuals who regularly experience cooperation, and who regularly benefit from it, will tend to develop more cooperative heuristics that kick in as the default response in social situations. If you have intuitively acted cooperatively in the past, and those actions have benefited you, then the heuristic is reinforced. Alternatively, those who have been rewarded for selfish behaviour will tend to develop uncooperative heuristics. There is variation across the population for this intuitive side of our behaviour (some people are more selfish than others), but the social heuristic hypothesis predicts that deliberation will tend to favour self-interested behaviour in everyone.

Greene and colleagues found very clear evidence that our intuitive 'default' mode is selflessness, and their work has become highly influential. However, while some studies have confirmed their original findings, others have found the complete opposite, with reflection leading instead to more cooperation.[3] To try to make sense of the developing confusion an analysis of 21 separate studies replicating the original work was published in 2017.[4] This analysis showed 'essentially no difference in contributions between the time-pressure [intuitive] and forced-delay [reflection] conditions'. Their conclusion was that the original finding was the product of biases arising from excluding people who didn't meet the time constraint. For now, the jury is well and truly out on whether we are less self-interested when we make rapid decisions. What is interesting though is that, regardless of the validity of the social heuristic hypothesis, studies find a mix of self-interested and selfless behaviours. While we can argue about the balance between intuition and reflection, and the contributions of innate hardwiring, learning and experience, the fact remains that a tendency towards self-interest is an important component of our evolved behavioural repertoire. We need to remember that tendency as we consider the future.

The very real problem of the very abstract 'future us'

When we consider the future, and more specifically 'our' future, we tend not to think of 'future us' as being the same person as 'present us'. We tend to view the hypothetical future version of ourselves as someone else, which means that we also tend to view the future as though it is happening to someone else. This overlap between what psychologists call temporal distancing (the difference between now and times in the past or future) and social distancing (the social distance between you and other people) is hugely relevant when it comes to thinking about how we might plan for the future. The sort of planning we need to do requires us to take decisions not just for future versions of ourselves but for future versions of everyone else, and indeed for future versions of people not yet in existence and their future offspring. That is an awful lot of temporal and social distance.

An interesting but not especially satisfying piece of evidence that we view our future-selves as different people comes literally from observations of how we view our future-self. When picturing future events, we tend to see our future-self from the perspective of an external observer, while we perceive our present-self first hand and experientially, from within. A much more satisfying and solid evidence base though comes from a series of experiments that were specifically designed to investigate this shift in perspective when it comes to the future.

One experiment involved a disgusting drink. Participants, in this case students, were asked to make either real or hypothetical decisions about how much of the disgusting drink should be drunk by themselves or another participant, either in the present or in the future. The set-up was that the drink was to be consumed as part of a fictitious scientific experiment about mood and perception. You cannot experience your present self (or your 'on-going self' as the experimenters term it) hypothetically and so the tendency is for the on-going self to focus attention on immediate subjective experiences. The prediction was therefore that participants making real decisions would choose to allocate more disgusting drink for their future-selves and for other people's future-selves than to themselves in the present. To add some real-world jeopardy to the future-based decisions, the students were told that they would have to drink the future disgusting drink (water, ketchup and soy sauce) early in the next semester and that they would lose academic credit if they didn't. Participants making hypothetical decisions were told to imagine themselves allocating the disgusting drink to the various different parties, including themselves. Sadly for the casual observer, in neither the real nor the hypothetical case, and neither in the present nor the future, did anyone actually end up drinking the disgusting drink.

In the hypothetical condition there were no differences between the amount of disgusting drink allocated to present and future selves or to other people. It seems that when it comes to hypothetical situations, we don't really distinguish. When it came to making what the participants thought was a real decision, however, the experimenters found a very big difference. In that case, the amount allocated by participants for themselves to drink right now was less than half that allocated to their future-self. The amount participants allocated to other people right now was, interestingly, identical to the amount they allocated to their future-self. In other words, when making decisions we tend to treat 'self' far better than 'others' but we don't carry this preferential treatment forward into the future. The disgusting drink experiment shows us that our priorities are firmly fixed in the order self first, then future-self and everybody else (regardless of when they exist).

Another experiment confirms this present–future bias. Again the experiment used students, this time asking them to volunteer time for a hypothetical peer-mentoring and tutoring programme initiated by a researcher posing as a fellow student. The experiment took place during the run-up to stressful and important mid-term examinations. Participants were divided into four groups:

1. self-present ('how much time would you give up this week?');
2. self-future simple ('how much time would you give up during the next midterm period?');
3. other ('how much time do you think other students could spare?'); and
4. self-future/same feelings ('how much time would you give up during the next midterm period *given that then you will be feeling exactly as stressed and anxious as you are right now?*').

The last group, self-future/same feelings were additionally told to remember that they would have about the same amount of work that they do right now, that they would still have the same concerns about their work–life balance and that they were going to be the same person that they are right now. The findings were very clear. Fewer minutes were volunteered for self-present (27 minutes) than for self-future simple (85 minutes) and for other people (a whacking 120 minutes). Once again the results show that we prioritise ourselves over all other people, including future versions of ourselves. What is really interesting is what happened when the experimenter took pains to outline to participants what 'future them' might be like. When participants were told that future-them was basically just present-them, they began to treat future-them much more like present-them, offering far fewer minutes (46 minutes).[1] The lesson here is that if we are reminded that we are going to be the same people in the future, with the same experiences, anxieties and fears, then we start to treat future us rather better. What we don't do is treat other people any better, now or in the future.

Other experiments, looking for example at receiving junk email on behalf of someone raising money for charity or examining the amount of prize money we take now or in the future, confirm the same basic findings. We discount future versions of ourselves and we prioritise 'present-self' over anybody else. As disappointing as this tendency might be, it cannot really come as any surprise. Just as in other animals, cognitive structures in our brains have evolved to allow us to function as here-and-now creatures that ultimately value self-preservation.

Of all our evolved tendencies, it is perhaps our attitude towards the future that might prove to be greatest mismatch in terms of impact. It is also the mismatch that might be the easiest to overcome. What the mentoring experiment shows us is that when we can engage our brain and think about the future, when we are given a chance to imagine future-us as a person identical in subjective experience to present-us, we start to treat future-us a little better. Whether that opportunity to think makes us less selfish towards other people (both now and in the future) is open to debate, but as we've seen, some experiments do suggest that this might be the case. There might just be some hope for us after all.

Saving ourselves

We make hundreds of decisions every day that have an impact on us and others, and on future versions of everyone. We have seen in previous chapters that taking an evolutionary approach towards understanding the mismatches that underlie some of our modern-world problems can give us insight into how we deal with problems like obesity or addiction. The big difference between the issues covered in earlier chapters and the ones we are considering now is that evolved tendencies to lead humans towards conditions like obesity or addiction can cause major societal problems, but they are not existentially threatening for humans overall. They harm many of us and they kill some of us but they don't affect us all. That is not necessarily the case with the selfish, future-discounting behaviours that we see currently degrading the planet. So, can considering our evolved tendencies in the way we think about other people and the future give us insights into how we might tackle the biggest problems that we face? I think it could.

At the moment, much of the environmental rhetoric is focused around blame, and much of that blame is levelled at countries with rapidly developing economies underpinned by models of power generation and the types of industries that tend to pollute (China, for example). That, of course, was the same in developed countries like the UK and USA not that long ago. It is nearly always politically expedient to blame others (remember social identity theory from Chapter 9?) but this just reinforces the idea of 'them and us'. It is a toxic approach that plays to our basic individual and group-level selfishness. We need to understand that globalisation is not just a way to get cheap plastic shoes from Indonesia; with the right mind-set, it is a perfect opportunity to overcome our individualistic and 'tribal' tendencies. As tentative and guarded a conclusion as it might be, we have seen that the more we see other people as part of our group, as 'us', the better we might treat them. It sounds cheesy and trite, but evolutionary insights suggest we need to work on ways to bring people together. Maybe John Lennon was on to something.

Then, of course, there is the fact that we tend to see future outcomes for future-us as being events that happen to someone else. This is a fundamental problem if our goal is to change behaviour today for the benefit of future generations. It is a problem that is made far worse by the fact that, like the disgusting drink in the future-self experiment, the medicine is going to be hard to swallow: we must consume less; we must pollute less; we must change our diets; we have to fly less, drive less and so on. Future us will have a lifestyle that is rather different, and less appealing, than present us. Perhaps to make us drink the medicine we need to play to our evolved foibles? A sensible first step would certainly be reinforcing, regularly and forcefully, the point that 'future us' is still 'us'. Future us will have the same outlook, feelings, hopes and fears as we do right now. Maybe the time has come to accept our limitations and play to our weaknesses? If so, then a better understanding of our evolutionary heritage might just give us some of the insights we need to save ourselves from the most important mismatch of all, and ensure that future generations can thrive on Earth.

References

Chapter 1

1 Gustafsson, A. and Lindenfors, P. 2009. Latitudinal patterns in human stature and sexual stature dimorphism. *Annals of Human Biology* 36: 74–87.
2 Grey, R. 2016. Cave fires and rhino skull used in Neanderthal burial rituals. *New Scientist*, 1 October 2016: 3093.

Chapter 2

1 Coronil, F., Fix, A.G., Pels, P., Briggs, C.L., Mantini-Briggs, C.E., Hames, R., Lindee, S. and Ramos, A.R. 2001. Perspectives on Tierney's Darkness in El Dorado. *Current Anthropology* 42: 265–76.
2 The letter and some additional commentary is available at people. maths.ox.ac.uk/trefethen/bmi.html.
3 Detail on the increase in diabetes globally can be found at www.who.int/news-room/fact-sheets/detail/diabetes.
4 For example, Vogli, R.D., Kouvonen, A., Elovainio, M. and Marmot, M. 2014. Economic globalization, inequality and body mass index: a cross-national analysis of 127 countries. *Critical Public Health* 24: 7–21.
5 Prentice, A.M. 2005. The emerging epidemic of obesity in developing countries. *International Journal of Epidemiology* 35: 93–9.
6 Prentice, A. and Webb, F. 2005. Obesity amidst poverty. *International Journal of Epidemiology* 35: 24–30.
7 Krishnan, M., Major, T.J., Topless, R.K., Dewes, O., Yu, L., Thompson, J.M., McCowan, L., de Zoysa, J., Stamp, L.K., Dalbeth, N. and Hindmarsh, J.H. 2018. Discordant association of the CREBRF rs373863828 A allele with increased BMI and protection from type 2 diabetes in Māori and Pacific (Polynesian) people living in Aotearoa/New Zealand. *Diabetologia* 61: 1603–13.
8 Speakman, J.R., 2008. Thrifty genes for obesity, an attractive but flawed idea, and an alternative perspective: the 'drifty gene' hypothesis. *International Journal of Obesity* 32: 1611.
9 Post, J.D. 1977. *The last great subsistence crisis in the Western World.* Johns Hopkins University Press, Baltimore.

10 Helmchen, L.A. and Henderson, R.M. 2004. Changes in the distribution of body mass index of white United States men, 1890–2000. *Annals of Human Biology* 31: 174–81.

11 Prentice, A.M., Hennig, B.J. and Fulford, A.J. 2008. Evolutionary origins of the obesity epidemic: natural selection of thrifty genes or genetic drift following predation release? *International Journal of Obesity* 32: 1607.

12 Wang, G. and Speakman, J.R. 2016. Analysis of positive selection at single nucleotide polymorphisms associated with body mass index does not support the 'thrifty gene' hypothesis. *Cell Metabolism* 24: 531–41.

13 Speakman, J.R. 2018. The evolution of body fatness: trading off disease and predation risk. *Journal of Experimental Biology* 221: p.jeb167254.

14 Speakman, J.R. 2006. Thrifty genes for obesity and the metabolic syndrome – time to call off the search? *Diabetes and Vascular Disease Research* 3: 7–11.

15 Cunningham, E. 2012. Are diets from paleolithic times relevant today? *Journal of the Academy of Nutrition and Dietetics* 112: 1296.

16 Crittenden, A.N. 2011. The importance of honey consumption in human evolution. *Food and Foodways* 19: 257–73.

17 Schoeninger, M.J. 2014. Stable isotope analyses and the evolution of human diets. *Annual Review of Anthropology* 43: 413–30.

18 Eaton, S.B. and Konner, M. 1985. Paleolithic nutrition: a consideration of its nature and current implications. *New England Journal of Medicine* 312: 283–9.

19 Mellberg, C., Sandberg, S., Ryberg, M., Eriksson, M., Brage, S., Larsson, C., Olsson, T. and Lindahl, B. 2014. Long-term effects of a Palaeolithic-type diet in obese postmenopausal women: a 2-year randomized trial. *European Journal of Clinical Nutrition* 68: 350.

20 Genoni, A., Lo, J., Lyons-Wall, P. and Devine, A. 2016. Compliance, palatability and feasibility of paleolithic and Australian guide to healthy eating diets in healthy women: A 4-week dietary intervention. *Nutrients* 8: 481.

Chapter 3

1 Latham, K.J. 2013. Human health and the Neolithic revolution: an overview of impacts of the agricultural transition on oral health, epidemiology, and the human body. *Nebraska Anthropologist* 28: 95–102.

2 Balk, E.M., Adam, G.P., Langberg, V.N., Earley, A., Clark, P., Ebeling, P.R., Mithal, A., Rizzoli, R., Zerbini, C.A.F., Pierroz, D.D. and Dawson-Hughes, B. 2017. Global dietary calcium intake among adults: a systematic review. *Osteoporosis International* 28: 3315–24.

3 Del Valle, H.B., Yaktine, A.L., Taylor, C.L. and Ross, A.C. (eds). 2011. *Dietary reference intakes for calcium and vitamin D*. National Academies Press.

4 Fuller, F., Beghin, J. and Rozelle, S. 2007. Consumption of dairy products in urban China: results from Beijing, Shangai and Guangzhou. *Australian Journal of Agricultural and Resource Economics* 51: 459–74.

5 For example, www.livekindly.com/chinas-growing-milk-consumption-global-concern.

6 Chen, P., Li, Z. and Hu, Y. 2016. Prevalence of osteoporosis in China: a meta-analysis and systematic review. *BMC Public Health* 16: 1039.

7 See commentary via the journal *Science* at www.sciencemag.org/news/2014/04/how-sheep-became-livestock.

8 Gerbault, P., Liebert, A., Itan, Y., Powell, A., Currat, M., Burger, J., Swallow, D.M. and Thomas, M.G. 2011. Evolution of lactase persistence: an example of human niche construction. *Philosophical Transactions of the Royal Society B: Biological Sciences* 366: 863–77.

9 Itan, Y., Jones, B.L., Ingram, C.J., Swallow, D.M. and Thomas, M.G. 2010. A worldwide correlation of lactase persistence phenotype and genotypes. *BMC Evolutionary Biology* 10: 36.

10 Bayoumi, R.A.L., Flatz, S.D., Kühnau, W. and Flatz, G. 1982. Beja and Nilotes: nomadic pastoralist groups in the Sudan with opposite distributions of the adult lactase phenotypes. *American Journal of Physical Anthropology* 58: 173–8.

11 Itan, Y., Powell, A., Beaumont, M.A., Burger, J. and Thomas, M.G. 2009. The origins of lactase persistence in Europe. *PLoS Computational Biology* 5: p.e1000491.

12 Bersaglieri, T., Sabeti, P.C., Patterson, N., Vanderploeg, T., Schaffner, S.F., Drake, J.A., Rhodes, M., Reich, D.E. and Hirschhorn, J.N. 2004. Genetic signatures of strong recent positive selection at the lactase gene. *The American Journal of Human Genetics* 74: 1111–20.

13 Lokki, A.I., Järvelä, I., Israelsson, E., Maiga, B., Troye-Blomberg, M., Dolo, A., Doumbo, O.K., Meri, S. and Holmberg, V. 2011. Lactase persistence genotypes and malaria susceptibility in Fulani of Mali. *Malaria Journal* 10: 9.

14 Rohrer, F. 2007. China drinks its milk. *BBC News* news.bbc.
    co.uk/1/hi/6934709.stm.
15 McClure, S.B., Magill, C., Podrug, E., Moore, A.M., Harper, T.K.,
    Culleton, B.J., Kennett, D.J. and Freeman, K.H. 2018. Fatty acid
    specific δ13C values reveal earliest Mediterranean cheese
    production 7,200 years ago. *PloS One* 13: p.e0202807.
16 Kropf, N.P. and Kelley, S. 2017. Why more grandparents are raising
    their grandchildren. *The Conversation* theconversation.com/
    why-more-grandparents-are-raising-their-grandchildren-83543.
17 De Munter, J.S., Hu, F.B., Spiegelman, D., Franz, M. and van Dam,
    R.M. 2007. Whole grain, bran, and germ intake and risk of type 2
    diabetes: a prospective cohort study and systematic review. *PLoS
    Medicine* 4: p.e261.
18 Jacobs Jr., D.R., Marquart, L., Slavin, J. and Kushi, L.H. 1998.
    Whole-grain intake and cancer: An expanded review and
    meta-analysis. *Nutrition and Cancer* 30: 85–96.
19 Ludvigsson, J.F., Leffler, D.A., Bai, J.C., Biagi, F., Fasano, A., Green,
    P.H., Hadjivassiliou, M., Kaukinen, K., Kelly, C.P., Leonard, J.N.
    and Lundin, K.E.A. 2013. The Oslo definitions for coeliac disease
    and related terms. *Gut* 62: 43–52.
20 Lohi, S., Mustalahti, K., Kaukinen, K., Laurila, K., Collin, P.,
    Rissanen, H., Lohi, O., Bravi, E., Gasparin, M., Reunanen, A. and
    Mäki, M. 2007. Increasing prevalence of coeliac disease over time.
    Alimentary Pharmacology and Therapeutics 26: 1217–25.
21 Kaukinen, K., Partanen, J., Mäki, M. and Collin, P. 2002.
    HLA-DQ typing in the diagnosis of celiac disease. *The American
    Journal of Gastroenterology* 97: 695–9.
22 Molina-Infante, J., Santolaria, S., Sanders, D.S. and Fernández-
    Bañares, F. 2015. Systematic review: noncoeliac gluten sensitivity.
    Alimentary Pharmacology and Therapeutics 41: 807–20.
23 Mansueto, P., Seidita, A., D'Alcamo, A. and Carroccio, A. 2014.
    Non-celiac gluten sensitivity: literature review. *Journal of the
    American College of Nutrition* 33: 39–54.
24 Lionetti, E., Gatti, S., Pulvirenti, A. and Catassi, C. 2015. Celiac
    disease from a global perspective. *Best practice & research Clinical
    Gastroenterology* 29: 365–79.
25 Morrell, K. and Melby, M.K. 2017. Celiac Disease: The Evolutionary
    Paradox. *International Journal of Celiac Disease* 5: 86–94.
26 Augusto, D.G. and Petzl-Erler, M.L. 2015. KIR and HLA under
    pressure: evidences of coevolution across worldwide populations.
    Human Genetics 134: 929–40.

27 Lionetti, E. and Catassi, C. 2014. Co-localization of gluten
    consumption and HLA-DQ2 and -DQ8 genotypes, a clue to the
    history of celiac disease. *Digestive and Liver Disease* 46: 1057–63.
28 Ivarsson, A. 2005. The Swedish epidemic of coeliac disease
    explored using an epidemiological approach – some lessons to
    be learnt. *Best practice & research Clinical Gastroenterology* 19:
    425–40.
29 Marasco, G., Di Biase, A.R., Schiumerini, R., Eusebi, L.H.,
    Iughetti, L., Ravaioli, F., Scaioli, E., Colecchia, A. and Festi, D.
    2016. Gut microbiota and celiac disease. *Digestive Diseases and
    Sciences* 61: 1461–72.

Chapter 4

1 Sender, R., Fuchs, S. and Milo, R. 2016. Revised estimates for the
   number of human and bacteria cells in the body. *PLoS Biology* 14:
   p.e1002533.
2 Yong, E. 2016. You're probably not mostly microbes. *The Atlantic*
   8 January 2016. www.theatlantic.com/science/archive/2016/01/
   youre-probably-not-mostly-microbes/423228/?dom=pscau&src
   =syn.
3 Clemente, J.C., Manasson, J. and Scher, J.U. 2018. The role of the
   gut microbiome in systemic inflammatory disease. *British Medical
   Journal* 360: p.j5145.
4 Wilson, J.C., Furlano, R.I., Jick, S.S. and Meier, C.R. 2015.
   Inflammatory bowel disease and the risk of autoimmune diseases.
   Journal of Crohn's and Colitis 10: 186–93.
5 Park, J.S., Lee, E.J., Lee, J.C., Kim, W.K. and Kim, H.S. 2007.
   Anti-inflammatory effects of short chain fatty acids in IFN-γ-
   stimulated RAW 264.7 murine macrophage cells: Involvement of
   NF-κB and ERK signaling pathways. *International
   Immunopharmacology* 7: 70–7.
6 Quoted in www.sciencedaily.com/releases/2017/10/171020125752.
   htm
7 For more information on MS see www.mayoclinic.org/diseases-
   conditions/multiple-sclerosis/symptoms-causes/syc-20350269.
8 Planas, R., Santos, R., Tomas-Ojer, P., Cruciani, C., Lutterotti, A.,
   Faigle, W., Schaeren-Wiemers, N., Espejo, C., Eixarch, H., Pinilla,
   C. and Martin, R. 2018. GDP-l-fucose synthase is a CD4+ T cell-
   specific autoantigen in DRB3★ 02: 02 patients with multiple
   sclerosis. *Science Translational Medicine*, 10: p.eaat4301.

9   Pennisi, E. 2019. Evidence mounts that gut bacteria can influence
    mood, prevent depression. *Science* 4 February 2019: www.
    sciencemag.org/news/2019/02/evidence-mounts-gut-bacteria-
    can-influence-mood-prevent-depression.
10 See editorial in *Nature* 566(7), 2019: www.nature.com/articles/
    d41586-019-00483-5.
11 Xu, Z. and Knight, R. 2015. Dietary effects on human gut
    microbiome diversity. *British Journal of Nutrition* 113: S1–S5.
12 Makki, K., Deehan, E.C., Walter, J. and Bäckhed, F. 2018. The
    impact of dietary fiber on gut microbiota in host health and
    disease. *Cell Host & Microbe* 23: 705–15.
13 Sonnenburg, E.D. and Sonnenburg, J.L. 2014. Starving our
    microbial self: the deleterious consequences of a diet deficient
    in microbiota-accessible carbohydrates. *Cell Metabolism* 20:
    779–86.
14 Xu, L., Lochhead, P., Ko, Y., Claggett, B., Leong, R.W. and
    Ananthakrishnan, A.N. 2017. Systematic review with meta-
    analysis: breastfeeding and the risk of Crohn's disease and
    ulcerative colitis. *Alimentary Pharmacology and Therapeutics* 46:
    780–9.
15 Pannaraj, P.S., Li, F., Cerini, C., Bender, J.M., Yang, S., Rollie, A.,
    Adisetiyo, H., Zabih, S., Lincez, P.J., Bittinger, K. and Bailey, A.
    2017. Association between breast milk bacterial communities and
    establishment and development of the infant gut microbiome.
    JAMA Pediatrics 171: 647–54.
16 Pawankar, R. 2014. Allergic diseases and asthma: a global public
    health concern and a call to action. *World Allergy Organization
    Journal* 7: 1–3.
17 Lundbäck, B., Backman, H., Lötvall, J. and Rönmark, E. 2016.
    Is asthma prevalence still increasing? *Expert Review of Respiratory
    Medicine* 10: 39–51.
18 Strachan, D. 1989. Hay fever, hygiene and household size. *British
    Medical Journal* 299: 1259–60.
19 Strachan, D. 2000. Family size, infection and atopy: the first
    decade of the 'hygiene hypothesis'. *Thorax* 55: S2–S10.
20 Stanwell-Smith, R., Bloomfield, S.F. and Rook, G.A.W. 2012.
    The hygiene hypothesis and its implications for home hygiene,
    lifestyle and public health. In *International Scientific Forum on Home
    Hygiene*: www.ifh-homehygiene.org/reviews-hygienehypothesis
    (an excellent and very readable review).

Chapter 5

1   Trichopoulos, D., Zavitsanos, X., Katsouyanni, K., Tzonou, A. and
    Dalla-Vorgia, P. 1983. Psychological stress and fatal heart attack: the
    Athens (1981) earthquake natural experiment. *The Lancet* 321: 441–4.

2   Nesse, R.M. and Young, E.A. 2000. Evolutionary origins and
    functions of the stress response. *Encyclopedia of Stress* 2: 79–84.

3   A summary of Ishigami's work can be read at www.brainimmune.
    com/mental-state-and-tuberculosis-tohru-ishigami-1918.

4   This and other advice is given in the JAMA Article 'Acute
    emotional stress and the heart', available at jamanetwork.com/
    journals/jama/fullarticle/208031.

5   A summary, with links to primary research, is available at www.
    healthline.com/health-news/millennial-depression-on-the-
    rise#Millennial-who?

6   Horwitz, A.V. and Wakefield, J.C. 2007. *The loss of sadness: How
    psychiatry transformed normal sorrow into depressive disorder.* Oxford
    University Press, Oxford.

7   Hidaka, B.H. 2012. Depression as a disease of modernity.
    explanations for increasing prevalence. *Journal of Affective Disorders*
    140: 205–14.

8   Breggin, P.R. 2015. The biological evolution of guilt, shame and
    anxiety: A new theory of negative legacy emotions. *Medical
    Hypotheses* 85: 17–24.

9   Statistics on anxiety and an overview of the many different forms
    of anxiety now recognised can be found at adaa.org/about-adaa/
    press-room/facts-statistics.

10 The Cancer Research United Kingdom page can be found at
    www.cancerresearchuk.org/about-cancer/causes-of-cancer/
    cancer-controversies/stress.

11 More information is available in the article 'These things don't
    cause cancer, but people think they do' by Yasemin Saplakoglu,
    2017, available at www.livescience.com/62412-widespread-
    cancer-myths.html.

12 Stein, D.J., Newman, T.K., Savitz, J. and Ramesar, R. 2006. Warriors
    versus worriers: the role of COMT gene variants. *CNS Spectrums*
    11: 745–8.

13 Gutleb, D.R., Roos, C., Noll, A., Ostner, J. and Schülke, O. 2018.
    COMT Val158Met moderates the link between rank and
    aggression in a non-human primate. *Genes, Brain and Behavior* 17:
    p.e12443.

Chapter 6

1 The recent problems of Myspace are detailed by Jon Bordkin in an article entitled 'Myspace apparently lost 12 years' worth of music, and almost no one noticed' available at arstechnica.com/information-technology/2019/03/myspace-apparently-lost-12-years-worth-of-music-and-almost-no-one-noticed.

2 Valkenburg, P.M., Peter, J. and Schouten, A.P. 2006. Friend networking sites and their relationship to adolescents' well-being and social self-esteem. *CyberPsychology and Behavior* 9: 584–90.

3 Wang, R., Yang, F. and Haigh, M.M. 2017. Let me take a selfie: Exploring the psychological effects of posting and viewing selfies and groupies on social media. *Telematics and Informatics* 34: 274–83.

4 Davila, J., Hershenberg, R., Feinstein, B.A., Gorman, K., Bhatia, V. and Starr, L.R. 2012. Frequency and quality of social networking among young adults: Associations with depressive symptoms, rumination, and corumination. *Psychology of Popular Media Culture* 1: 72.

5 Feinstein, B.A., Bhatia, V. and Davila, J. 2014. Rumination mediates the association between cyber-victimization and depressive symptoms. *Journal of Interpersonal Violence* 29: 1732–46.

6 Dobrean, A. and Păsărelu, C.R. 2016. Impact of Social Media on Social Anxiety: A Systematic. *New developments in Anxiety Disorders*: 129.

7 A general discussion of the effects of social media, and the evidence or otherwise of its ill effects, can be read at www.bbc.com/future/story/20180104-is-social-media-bad-for-you-the-evidence-and-the-unknowns.

8 Arampatzi, E., Burger, M.J. and Novik, N. 2018. Social network sites, individual social capital and happiness. *Journal of Happiness Studies* 19: 99–122.

9 Lakhiani, V. 2018. Science-based reasons why friends make us happier. *Mind Valley* 17 January 2018: blog.mindvalley.com/why-friends-make-us-happier.

10 Matsumoto-Oda, A. and Oda, R. 1998. Changes in the activity budget of cycling female chimpanzees. *American Journal of Primatology* 46: 157–66.

11 Takano, M. 2018. Two types of social grooming methods depending on the trade-off between the number and strength of social relationships. *Royal Society Open Science* 5: p.180148.

12 Brondino, N., Fusar-Poli, L. and Politi, P. 2017. Something to talk about: gossip increases oxytocin levels in a near real-life situation. *Psychoneuroendocrinology* 77: 218–24.

13 Dunbar, R.I.M. 1993. Co-Evolution of neocortex size, group size and language I in humans. *Behavioral and Brain Sciences* 16: 681–735.

14 Dunbar, R.I.M. 1992. Neocortex size as a constraint on group size in primates. *Journal of Human Evolution* 22: 469–93.

15 Buettner, R. 2017. Getting a job via career-oriented social networking markets. *Electronic Markets* 27: 371–85.

16 The W.L. Gore model is described in www.ideaconnection.com/interviews/00012-The-Tipping-Point.html.

17 Healy, S.D. and Rowe, C. 2006. A critique of comparative studies of brain size. *Proceedings of the Royal Society B: Biological Sciences* 274: 453–64.

18 Dunbar, R.I.M. 2012. Social cognition on the Internet: testing constraints on social network size. *Philosophical Transactions of the Royal Society B: Biological Sciences* 367: 2192–201.

19 McCarty, C., Killworth, P.D., Bernard, H.R., Johnsen, E.C. and Shelley, G.A. 2001. Comparing two methods for estimating network size. *Human Organization* 60: 28–39.

20 Zhou, W.X., Sornette, D., Hill, R.A. and Dunbar, R.I. 2005. Discrete hierarchical organization of social group sizes. *Proceedings of the Royal Society B: Biological Sciences* 272: 439–44.

21 Dunbar, R.I.M. 2014. Mind the gap: or why humans aren't just great apes. In *Lucy to Language: The Benchmark Papers*: 3–18. Oxford University Press, Oxford.

22 Wellman, B. 2012. Is Dunbar's number up? *British Journal of Psychology* 103: 174–6.

23 Facebook statistics available via www.omnicoreagency.com/facebook-statistics.

Chapter 7

1 Ferguson, C.J. and Beaver, K.M. 2009. Natural born killers: The genetic origins of extreme violence. *Aggression and Violent Behavior* 14: 286–94.

2 Gómez, J.M., Verdú, M., González-Megías, A. and Méndez, M. 2016. The phylogenetic roots of human lethal violence. *Nature* 538: 233–7.

3 GBH, grievous bodily harm, is the most severe definition of assault in English criminal law and requires wounding or other severe injury to be inflicted with intent.

4 Frisell, T., Lichtenstein, P. and Långström, N. 2011. Violent crime runs in families: a total population study of 12.5 million individuals. *Psychological Medicine* 41: 97–105.

5 Zhang-James, Y., Fernàndez-Castillo, N., Hess, J.L., Malki, K., Glatt, S.J., Cormand, B. and Faraone, S.V. 2018. An integrated analysis of genes and functional pathways for aggression in human and rodent models. *Molecular Psychiatry* 1.

6 The BBC provides an overview of 'one punch deaths' in an article available at www.bbc.co.uk/news/uk-38992393.

7 An account of the exchange between Carrier and Fish is provided by Amina Khan in the *LA Times*: www.latimes.com/science/sciencenow/la-sci-sn-human-fist-punching-evolution-males--20151021-story.html.

8 Geoffrey Mohan discusses this hypothesis and its reception in the *LA Times*: www.latimes.com/science/sciencenow/la-sci-sn-face-fit-for-punches-20140609-story.html.

9 Nickle, D.C. and Goncharoff, L.M. 2013. Human fist evolution: a critique. *Journal of Experimental Biology* 216: 2359–60.

10 Oka, R.C., Kissel, M., Golitko, M., Sheridan, S.G., Kim, N.C. and Fuentes, A. 2017. Population is the main driver of war group size and conflict casualties. *Proceedings of the National Academy of Sciences* 114: E11101–E11110.

11 Michael Price provides an excellent overview of this research in *Science*: www.sciencemag.org/news/2017/12/why-human-society-isn-t-more-or-less-violent-past.

12 A 'reality check' of the situation in the United Kingdom is provided by BBC Reality Check at www.bbc.co.uk/news/uk-44397532.

13 Media Violence Commission, International Society for Research on Aggression (ISRA). Report of the Media Violence Commission. Aggressive Behavior 38: 335–41.

14 Discussed in www.theguardian.com/voluntary-sector-network/2014/sep/29/poverty-porn-charity-adverts-emotional-fundraising.

15 A nice overview of some of this research is given in www.telegraph.co.uk/news/science/11087683/Watching-violent-films-does-make-people-more-aggressive-study-shows.html.

Chapter 8

1 A medically reviewed summary of the costs of drugs is given at www.verywellmind.com/what-are-the-costs-of-drug-abuse-to-society-63037.

2 The Institute of Alcohol Studies is an excellent resource for alcohol-related facts and figures: www.ias.org.uk/Alcohol-knowledge-centre/Economic-impacts/Factsheets/Estimates-of-the-cost-of-alcohol.aspx.

3 Goodchild, M., Nargis, N. and d'Espaignet, E.T. 2018. Global economic cost of smoking-attributable diseases. *Tobacco Control* 27: 58–64.

4 www.drugabuse.gov/related-topics/trends-statistics/overdose-death-rates.

5 The National Institute on Drug Abuse provides an excellent overview of cocaine: www.drugabuse.gov/publications/teaching-packets/neurobiology-drug-addiction/section-iv-action-cocaine.

6 Kilts, C.D., Schweitzer, J.B., Quinn, C.K., Gross, R.E., Faber, T.L., Muhammad, F., Ely, T.D., Hoffman, J.M. and Drexler, K.P. 2001. Neural activity related to drug craving in cocaine addiction. *Archives of General Psychiatry* 58: 334–41.

7 Self, D.W. 1998. Neural substrates of drug craving and relapse in drug addiction. *Annals of Medicine* 30: 379–89.

8 Nutt, D., King, L.A., Saulsbury, W. and Blakemore, C. 2007. Development of a rational scale to assess the harm of drugs of potential misuse. *Lancet* 369: 1047–53.

9 Nutt, D. 2008. Equasy – an overlooked addiction with implications for the current debate on drug harms. *Journal of Psychopharmacology* 23: 3–5.

10 The debacle is described in www.theguardian.com/politics/2008/may/08/drugspolicy.drugsandalcohol.

11 Plowman, T. and Rivier, L. 1983. Cocaine and cinnamoylcocaine content of *Erythroxylum* species. *Annals of Botany* 51: 641–59.

12 Additional information on the manufacture of cocaine can be found at www.recovery.org/cocaines/how-made/.

13 www.bbc.co.uk/newsbeat/article/40015726/cocaine-is-getting-stronger-drug-experts-warn.

14 More information on coca leaf chewing can be found in the United Nations Office on Drugs and Crime Bulletin: www.unodc.org/unodc/en/data-and-analysis/bulletin/bulletin_1952-01-01_2_page009.html.

15 Jikomes, N. and Zoorob, M. 2018. The cannabinoid content of legal cannabis in Washington State varies systematically across testing facilities and popular consumer products. *Scientific Reports* 8: 4519.
16 Steigerwald, S., Wong, P.O., Khorasani, A. and Keyhani, S. 2018. The form and content of cannabis products in the United States. *Journal of General Internal Medicine* 33: 1426–8.
17 Gage, S.H. 2019. Cannabis and psychosis: triangulating the evidence. *The Lancet Psychiatry* 6: 364–5.
18 Sun, A.J. and Eisenberg, M.L. 2017. Association between marijuana use and sexual frequency in the United States: A population-based study. *The Journal of Sexual Medicine* 14: 1342–7.
19 Zemishlany, Z., Aizenberg, D. and Weizman, A. 2001. Subjective effects of MDMA ('Ecstasy') on human sexual function. *European Psychiatry* 16: 127–30.
20 Boileau, I., Assaad, J.M., Pihl, R.O., Benkelfat, C., Leyton, M., Diksic, M., Tremblay, R.E. and Dagher, A. 2003. Alcohol promotes dopamine release in the human nucleus accumbens. *Synapse* 49: 226–31.
21 Carrigan, M.A., Uryasev, O., Frye, C.B., Eckman, B.L., Myers, C.R., Hurley, T.D. and Benner, S.A. 2015. Hominids adapted to metabolize ethanol long before human-directed fermentation. *Proceedings of the National Academy of Sciences* 112: 458–63.
22 The State of Science Report on the Effects of Moderate Drinking by the National Institutes of Health and Department of Health and Human Services 2003 available at pubs.niaaa.nih.gov/publications/ModerateDrinking-03.htm.
23 This UK Government site tells you exactly how much you can expect to lose on gaming machines www.gamblingcommission.gov.uk/for-the-public/Safer-gambling/Consumer-guides/Machines-Fruit-machines-FOBTs/Gaming-machine-payouts-RTP.aspx.
24 Paulson, J., 2018. Hardwired for Risk: The Clinical Utility of Exploring Evolutionary Aspects of Gambling. *Journal of Gambling Issues* 40: 174–9.

Chapter 9

1 This, and more of Margaret Sullivan's thoughts on fake news, can be read at newslab.org/fake-news-has-lost-its-meaning-and-punch-posts-margaret-sullivan-says.

2 MacDonald, E. 2017. The fake news that sealed the fate of Antony and Cleopatra. *The Conversation* 13 January 2017: theconversation.com/the-fake-news-that-sealed-the-fate-of-antony-and-cleopatra-71287.

3 Soll, J. 2016. The long and brutal history of fake news. *Politico Magazine* 18 December 2016: www.politico.com/magazine/story/2016/12/fake-news-history-long-violent-214535.

4 Hill, R. and Flanagan, J. 2019. The Maternal–Infant Bond: Clarifying the Concept. *International Journal of Nursing Knowledge*: onlinelibrary.wiley.com/doi/pdf/10.1111/2047-3095.12235.

5 Kosfeld, M., Heinrichs, M., Zak, P.J., Fischbacher, U. and Fehr, E. 2005. Oxytocin increases trust in humans. *Nature* 435: 673.

6 Nave, G., Camerer, C. and McCullough, M. 2015. Does oxytocin increase trust in humans? A critical review of research. *Perspectives on Psychological Science* 10: 772–89.

7 Ide, J.S., Nedic, S., Wong, K.F., Strey, S.L., Lawson, E.A., Dickerson, B.C., Wald, L.L., La Camera, G. and Mujica-Parodi, L.R. 2018. Oxytocin attenuates trust as a subset of more general reinforcement learning, with altered reward circuit functional connectivity in males. *Neuroimage* 174. 35–43.

8 McNamara, J.M., Stephens, P.A., Dall, S.R. and Houston, A.I. 2008. Evolution of trust and trustworthiness: social awareness favours personality differences. *Proceedings of the Royal Society B: Biological Sciences* 276: 605–13.

9 Engelmann, J.M. and Herrmann, E. 2016. Chimpanzees trust their friends. *Current Biology* 26: 252–6.

10 Brosseau-Liard, P., Cassels, T. and Birch, S. 2014. You seem certain but you were wrong before: Developmental change in preschoolers' relative trust in accurate versus confident speakers. *PloS One 9*: p.e108308.

11 Talamas, S.N., Mavor, K.I. and Perrett, D.I. 2016. Blinded by beauty: Attractiveness bias and accurate perceptions of academic performance. *PloS One* 11: p.e0148284.

12 Lalumière, M.L., Harris, G.T. and Rice, M.E. 2001. Psychopathy and developmental instability. *Evolution and Human Behavior* 22: 75–92.

13 Graham, J. and Özener, B. 2016. Fluctuating asymmetry of human populations: a review. *Symmetry* 8: 154.

14 Haslam, S.A. and Reicher, S.D. 2017. 50 years of 'obedience to authority': From blind conformity to engaged followership. *Annual Review of Law and Social Science* 13: 59–78.

15 Jarret, C. 2017. New analysis suggests most Milgram participants
    realised the 'obedience experiments' were not really dangerous.
    The British Psychology Society Research Digest: digest.bps.org.
    uk/2017/12/12/interviews-with-milgram-participants-provide-
    little-support-for-the-contemporary-theory-of-engaged-
    followership/.

Chapter 10

1 Pronin, E., Olivola, C.Y. and Kennedy, K.A. 2008. Doing unto
    future selves as you would do unto others: Psychological distance
    and decision making. *Personality and Social Psychology Bulletin* 34:
    224–36.
2 Rand, D.G., Greene, J.D. and Nowak, M.A. 2012. Spontaneous
    giving and calculated greed. *Nature* 489: 427.
3 Lohse, J. 2016. Smart or selfish – When smart guys finish nice.
    Journal of Behavioral and Experimental Economics 64: 28–40.
4 Bouwmeester, S., Verkoeijen, P.P., Aczel, B., Barbosa, F., Bègue, L.,
    Brañas-Garza, P., Chmura, T.G., Cornelissen, G., Døssing, F.S.,
    Espín, A.M. and Evans, A.M. 2017. Registered replication report:
    Rand, Greene, and Nowak (2012). *Perspectives on Psychological
    Science* 12: 527–42.

Acknowledgements

I would like to thank the BBC Radio Science team for their tireless dedication to bringing science to life and for inadvertently inspiring me to write this book. I also thank the team at Bloomsbury Sigma Science for their support, especially Jim Martin, Anna MacDiarmid and Catherine Best. Thanks also to the team at Deborah McKenna, with special thanks to Borra Garson and Jan Croxon for their continuing support and patience.

Index